# Biometric Technology and Authentication

This book is an update to the first edition of this book. Some of the updates include newer images and an extensive review as to how both AWS and Microsoft Azure can be used to host a biometrics in the cloud infrastructure, which will be a hot topic going into 2025 and beyond. Finally, the last chapter of this book previews some of the latest trends for biometric technology going into 2025 and beyond.

**Ravi Das** is a technical writer in the cybersecurity realm. He also does cybersecurity consulting on the side through his private practice, Technical Writing Consulting, Inc. He holds the Certified in Cybersecurity Certification from the ISC(2).

# Biometric Technology and Authentication

## Biocryptography and Cloud-Based Architecture

### Second Edition

Ravi Das

CRC Press
Taylor & Francis Group
Boca Raton London New York

CRC Press is an imprint of the
Taylor & Francis Group, an **informa** business

Designed cover image: Shutterstock

Second edition published 2025
by CRC Press
2385 NW Executive Center Drive, Suite 320, Boca Raton FL 33431

and by CRC Press
4 Park Square, Milton Park, Abingdon, Oxon, OX14 4RN

*CRC Press is an imprint of Taylor & Francis Group, LLC*



First edition published by Taylor & Francis Group 2015



ISBN: 978-1-032-91072-7 (hbk)
ISBN: 978-0-367-67007-8 (pbk)
ISBN: 978-1-003-56214-6 (ebk)

DOI: 10.4324/9781003562146

Typeset in Sabon
by Newgen Publishing UK

This book is dedicated to my Lord and Savior, Jesus Christ, the Grand Designer of the Universe, and to my parents, Dr. Gopal Das and Mrs. Kunda Das, and to my loving cats, Fifi and Bubu, and guinea pig, Noodles.

This book is also dedicated to:
Richard and Gwynda Bowman
Jaya Chandra
Tim Auckley
Patricia Bornhofen
Ashish Das
Caylee Gibbons
Rory Maxfield
Caylee Gibbons

# Contents

# Acknowledgment

I would like to thank Ms. Gabrielle Williams, my editor, who made this book a reality.

Chapter 1

# Introduction to biometrics

If one were to critically examine biometric technology on a spectrum in comparison with the other security mechanisms that are available to us today, it would rank very high (if not the highest) in terms of curiosity, understanding, and social impacts. To illustrate this point, consider your everyday lifestyle.

You get up at a certain time to go to work, you perform your job functions according to the daily schedule, you return home, spend time with your family, and get ready yet once again for the next workday. Then there are, of course, the weekends that are available to us to get caught up on personal matters and chores around our household. From the moment that we get up, we are faced with security. When we take a shower, we can lock the bathroom door, and when we leave our place of residence, we make sure that the main doors are locked and made secure as well.

We unlock our cars, and when we arrive at our place of employment, we often have to prove our identity. This can be done merely by showing our identification badge, or if the business or organization is sophisticated enough as well, they might employ a smart card-based system.

With this technology, our personal information and data are stored in the memory bank of the smart card, and this is all confirmed by the smart card reader as we swipe it into the electro-magnetic slide at the turnstile.

When we arrive at our workstation, we have yet another fortitude of security to go through. This is primarily done by entering our username and password combination at our computer. This, of course, just gives us authority to access resources at the local level.

If we need to access confidential and proprietary information at a much more granular level (for instance, gaining access to a particular network drive), we then have to establish and enter a whole new suite of usernames and passwords.

This goes on the entire workday. If we have to conduct virtual meetings and the like, there are yet more usernames and passwords to be entered. Then as the workday comes to an end, we walk to our car, and unlock it. As

DOI: 10.4324/9781003562146-1

we proceed to step out and enter our respective homes, there is yet another layer of security we have to go through, which is unlocking the door lock.

But, if we choose to implement a multimodal security solution (such as an ADT burglar alarm), we have to quickly remember to disarm it.

Then, as the weekend fast approaches, we face yet another host of security to go through as we set out to conduct our objectives. For example, if we choose to fly to a particular destination, we will have to go through the security mechanisms at the airport, which will involve primarily walking through a magnetometer and having our driver's license or even passport scrutinized in order to confirm and validate our identity. Of course also, our shoes will have to be scanned as well to make sure that as well in order to ensure the security officials that we do not have a bomb on them.

But if we choose to stay around our locales for the weekend, we still face security measures to go through as well. For instance, if we go grocery shopping, if we choose to purchase our products with a check, we will have to establish our identity to the cashier by showing our driver's license yet once again, or even a state-issued identification card.

Or if we choose to attend a large-scale social event (such as that of a concert, sports festival, or other type or kind of get-together, which involves a lot of people), the chances are high that we will have to show our infamous driver's license yet once again and even walk through a magnetometer.

Now, these examples just illustrated the assumption that the average person is in a non-technology-related job. But what if this particular individual is in a career that relates to security technology? Obviously, then he or she will be involved in installing, deploying, configuring, and even troubleshooting these devices on a daily basis.

For example, this could involve having a direct interface with network security items, such as firewalls (which inspect for and forbid malformed data packets from entering into the network infrastructure of a place of business or organization), network intrusion devices (these types of technologies literally sniff out and detect threats that are transpiring in a particular network segment, isolate them, and mitigate the risk or threat), or even routers (these devices route the flow of data packets on a given network trunk securely and safely via the use of mathematically based routing tables).

The point of all this is, whether it is just the average person or even a certified and educated security professional, we do not even think twice about the security measures we face each and every single day. For example, as you enter your username and password into your employer's workstation, do you think twice about it (except for the frustration and anger, which it will cause if indeed you do forget your password and have to have it reset again – but this time with a much longer and more complex one)? For sure you do not; it just becomes a routine part of everyday professional life.

Or what about those days in which you go to a grocery store to pay by check and you are asked to show your state-issued identification card ... do you think twice about that? Probably not, other than the fact it is just a pure hassle and perhaps a waste of time, particularly if you are in a rush to be at your next destination.

Or what if you are attending a large social event such as the ones just described, do you ever think twice to ponder about the levels of security which are implemented there? Once again, probably not here either. The only thought that is transpiring in your mind is, why is so much security needed, and how soon can I get to my reserved spot?

In terms of the security professional, when dealing with the types and kinds of technologies just detailed, they do not even entertain a second thought about it. Why is this so? Well, of course, he or she deals with them on a daily basis.

In other words, to them, it is second nature, and no extra thought is needed or required. In reality, though, the only other thought that would enter his or her mind is if the layers of security that they have deployed and implemented will be enough to thwart off any cyber-related threats to their business or organization by which they are employed.

Then why is it that when we compare all of these typical scenarios with that of biometric technology, it gets the most level of scrutinization? When we register ourselves with a particular biometric system, it is a piece, or a snapshot of our individual selves which is being captured, in order to help confirm our identity at a later point in time.

For example, from the standpoint of physiologically based biometric technologies (which are also known simply as the "Physical Biometrics"), it could be an image of our fingerprint, vein pattern, face, eyes (in particular either the retina or the iris), hand shape, or even our voice that is being captured.

From the standpoint of behaviorally based biometric technologies (which are also known simply as behavioral biometrics), it could be mannerisms in the way we sign our name or even type upon a computer that could be captured.

If you take some time to think about it, no other type or kind of security technology does this – not even the firewalls, network intrusion devices, or even the routers. It is only biometric technology that does this. This is one of the major reasons why we are so intrigued by it, fascinated by it, or even scared by it. We have no control over those images that are being captured or even recorded upon us.

This merely refers to the physiological or behavioral image that is being captured by the biometric system; it is captured, but we have no idea as to how it will be processed, computed, and even stored into a database. It is this latter situation that causes one of the highest levels of apprehension.

For instance, we do not know under what methods the biometric templates (these are the mathematical files that are computed from the raw images) will be stored, how they will be protected from cyberattacks, or even worst yet, if they will be used maliciously by some third party, such as the Federal Government, when it comes to purposes of practicing high levels of law enforcement.

Also, when compared to the other types and kinds of security technologies that are available today (once again, the routers, the firewalls, and the network intrusion devices), there is a fundamental lack of understanding of what truly biometric technology is all about.

Of course, this same type of hypothesis could be very easily extended as well to the network security technologies just described, but yet once again, it is biometric technology that gets picked at over and over again.

Also, then there comes the issue of the societal impacts that biometric technology has upon the citizens of the world. Truth be told, many nations and their respective governments are actually quite accepting and even exuberated when it comes time to deploy and implement a biometrics-based infrastructure for their country, whether it is for the e-Passport, e-Voting, border security, a National ID Card scheme, or even to fortify the existing wide area network (WAN) or even a local area network (LAN).

This trend appears to be occurring quite heavily and rapidly in the developing nations. This is primarily due to the fact that the use of biometric technology gives the citizens of these developing nations a sense of self worth, fulfillment, and most importantly, knowing the fact that he or she is actually being recognized as a unique and single citizen in the eyes of their own government, especially when it comes time to receiving their particular allotted government entitlements and benefits.

But, in the developed nations, such as those of Europe, Australia, and the United States in particular, the societal impacts are much stronger and are also very negative. The primary reason for this is that as citizens of the United States, we are endowed with certain inalienable rights that are guaranteed to us by our very own Constitution.

For example, some of these include the rights to privacy, freedom, the ability to vote for whom we choose in the time of elections, and to pretty much lead our lives the way we want to as long as it does not intrude upon the rights or the welfare of our fellow citizens. Because of this fact, we can claim that using biometric technology intrudes upon our said rights and freedoms.

In other words, we can lay the foundation for the argument that biometric technology is a sheer intrusion upon our civil liberties and privacy rights. This stands in stark opposition to the citizens of the developing nations, where such freedoms and rights have no guarantee or place in those respective societies.

Thus, this is the primary reason for the social acceptance of biometric technology in these specific geographic regions being so high – because it helps to aid or give these citizens such cherished rights as we enjoy them here in the United States.

But, because we are already guaranteed all this by our Constitution, we can totally decry the use of biometric technology with a great margin of comfort, still knowing that there are certain mechanisms in place, which will guarantee us that no matter what, we will still be counted as unique and single citizens by our own Federal Government, and that no matter what events transpire in our lives (of course to a certain degree), we will still be able to claim and receive our entitlements and benefits. And if we do not receive them in a timely manner, there are certain legal recourses established for us by which we can still receive them, albeit at a later point in time.

As mentioned, although the traditional security technologies (which include those from the standpoint of both physical- and logical-based access) will continue to serve individuals, partnerships, corporations, organizations, and businesses worldwide, it will be biometric technology that will come to the forefront in the coming future. This is primarily due to a number of reasons, which include the following:

- In terms of logical access entry, the use of the username and password combination is also becoming counterproductive to any types and kinds of security measures. For example, corporations and businesses are now mandating the use of complex passwords that are very difficult for the end user to remember. This is so because passwords are now the main points of covert cyberattacks and can be easily guessed in just a matter of seconds with the right hacking technologies. And with much longer and more complex passwords, employees are now resorting to writing down these passwords on post it notes and literally attaching them to their workstation monitors. This can be easily seen and detected by others, especially by non-employees of the business or organization, such as contractors, and as a result, large scale cases of identity theft (ID Theft) can occur. Even worse, the costs associated with resetting these complex passwords are costly to reset to an information technology (IT) team, which can be as much as $300 per employee. For a small-scaled business or organization, these costs and expenses may not be as much, but as the employee size grows over time, these costs can become quite steep and absorbitant. As a result, other means are being looked at to replace the traditional username and password combination, and the use of biometric technology is receiving very close attention in this aspect.
- In terms of physical access entry, the traditional lock and key methods are often used. While there is no doubt that they provide a decent

means of securing a building or even an internal office setting, they can be very easily tampered with. For example, a padlock can be very easily cut off or picked at enough until it opens up. Even built-in door locks do not provide adequate protection; if a person has enough intent, he or she can even literally take a gun and simply blow it off in just a matter of seconds. Apart from this traditional lock and key approach, there are now much more sophisticated means in which to protect the physical premises of a building. For instance, smart card technology is now being used as well. Smart cards look just like a credit card – but with one major difference. A smart card actually consists of a specialized memory bank, and in this, all of the information and data about a card bearer can be stored here. So, for example, in order to confirm the identity of a particular individual, all he or she merely has to do is either flash their smart card (which is really their work-related identification card) in front of a smart card reader or swipe in a turnstile as he or she walks through it. If the information and data that are stored in the memory bank of the smart card match and correspond to what is stored in the employee database at the place of business or organization, then that particular employee is, of course, allowed entrance. If these credentials do not match up, then obviously, he or she is not allowed to gain access. But here too, there are inherent security weaknesses. For example, what if the employee loses their smart card-based ID badge? It can be very easily picked up by a person not affiliated with the business or organization, and unless the security guard is actually comparing the picture on the ID badge to the face of the person, this impostor could get inside a business or organization quite easily. Or worse yet, if the hacker is covert and sophisticated enough, he or she could make use of a network sniffer and actually intercept the information and data that reside upon the memory banks of the smart card. From here, the cyberattacker could then create a legitimate and authentic-looking smart card ID badge and import the intercepted information and data into the memory bank of this faked ID badge and gain easy and quick access into the business or organization. The primary reason why this can happen is that when smart cards are read by a particular smart card reader, the communications between the two are based upon a wireless protocol known specifically as "Near Field Communications", or "NFC" for short. This form of communication is unencrypted; thus, the information and data that reside in the memory bank of the smart card can be intercepted as plaintext, without any sort of deciphering required on the part of the hacker. Because of these security weaknesses with both the traditional lock and key as well as smart card technology – the use of the various biometric modalities is also being called upon quite heavily in this market segment as well.

- With regards to time and attendance applications, in most places of businesses and organizations, the use of the traditional time card is still used. Given the recent advancements in technology, although this method may seem to be very archaic as well as rudimentary, people are resistant to change and still continue to use the time card as the primary means in which to record the hours their employees have worked. But this method poses not only an administrative headache but also a huge security threat as well. For instance, with the former, the time as it is recorded by the time clock and punched to the respective time card has to be entered manually into a spreadsheet of sorts. This is not only time consuming and laborious, but many clerical-related errors can be introduced into the spreadsheet, and the cost it takes to hire the administrative staff to enter all of these hours in can be quite costly as well as expensive for any place of business or organization. Also, one must consider, what if the corporation is hit with either a man-made or natural disaster? What will happen to all of these time card records as well? Obviously, there is no easy way to electronically back them up, and if this particular scenario were to actually happen, once again, it would be very laborious and costly to scan in each and every time card as set forth by the security policies of the place of business or organization. In terms of the latter point, probably the greatest security threat is that of the phenomenon known as "Buddy Punching". For example, in this particular scenario, an employee calls in sick from work and, instead, has his or her friend (who is another employee of the business or organization) punch in and punch out their time card. So in other words, the "sick" employee still gets paid even when they were not actually "sick". This problem is probably not so much the case with a small-sized business, as each and every employee can be visible seen working and accounted for. But this is not the case with a much larger corporation. Given the much greater employee pool, each and every one cannot be accounted for, and their work rotations observed with the naked eye. Thus, in this regard, the use of biometric technology is also being looked at very heavily – not only to help automate and keep the administrative costs associated with time keeping down, but also to eliminate the security threat of "Buddy Punching" altogether.
- In terms of security documents, the use of the driver's license and even the paper passport has been used heavily in order to confirm the actual identity of a particular individual. The former is used mostly at the local and state levels, and the latter is used mostly as a means of

identification at the federal level (such as proving one's identity at the major international airports, or when an individual travels overseas). Although these types and kinds of security documents have literally stood the test of time, they too have a major security vulnerability. That is, given the advancements in technology, a driver's license or even the traditional paper passport can be quite easily spoofed and even replicated. Once again, the use of smart card technology has been used here as well to circumvent this particular security threat, but the information and data can also be easily intercepted here as well, primarily through the use of a network sniffer. Once again, it is the Near Field Communication wireless protocol that is being used and offers no means of encryption whatsoever. So, in order to prevent the threats of easy replication of these types and kinds of security documents, as well as the easy interception of the information and data that reside in the memory bank of a smart card, the use of biometric technology has also been called upon in this market segment as well.

Given these reasons now as to why biometric technology will now become one of the predominant security technologies to come out into the marketplace, both the technical and the social issues surrounding the use of biometric technology will still continue to persist and evolve even more. It is therefore very important to make an attempt to resolve these particular issues (both technical and social), so that the adoption rate of biometric technology can continue to grow in the developed nations (versus the developing nations), especially here in the United States.

But it is important to remember that the social and technical issues that surround biometric technology are complex ones; in other words, there is no easy solution to them. Perhaps one of the best ways the developed nations, like that of the United States, can learn how to actually embrace biometric technology and adopt it properly into mainstream society is to examine closely the developing nations (such as those in Africa, Asia, and Eastern Europe) that have taken up using the various biometric modalities that are now available.

For example, it has long baffled social science scholars as to why the developing nations around the world can quickly embrace new technologies such as the various biometric modalities, and why the developed nations of the world (in particular the United States) cannot. It really appears as a paradox, for example, the developing geographic regions really do not possess a strong enough IT infrastructure to support the most sophisticated means of security.

But for some reason or another, the governments of these nations are able to quickly and, to some degree, easily and efficiently create a national ID card system, an e-Passport infrastructure, very strong levels of border security, and even create and deploy an infrastructure to thwart off any cyberattacks in regard to the logical-based assets of that particular developing nation.

If you take a moment to think about it, you do not hear much of these developing nations as being a primary victim of a cyber-based attack or threat; rather, it is all targeted toward the developed nations.

Now take the example of the developed nations, in particular those of Australia, the United States, and those in Europe as well. Here, the most modern technological infrastructures are present, and one would think that it would be very easy to deploy a biometric system without any problems at all. But still, this problem persists.

There is huge resistance to not just deploying a biometrics-based infrastructure in any way, shape, or form, but there are also strong hesitancies to incorporate any type or kind of national ID card system, an e-Passport infrastructure, or even adopting non-biometric means in which to secure the critical IT infrastructure.

Now, when in comparison to the developing nations, it is the United States that has been one of the prime targets for cyberattacks and threats, mostly stemming from the developing nations, such as those of North Korea and Eastern Russia. Why do these sharp differences exist between the developing nations and the developed nations? Once again, it comes to the standard of living and the constitutional rights that we are afforded.

For example, in the developing nations, the standard of living is obviously very low when compared to that of the Western standards of living. Thus, the citizens of these nations yearn for every opportunity for a better and to be counted as citizens in the eyes of their own government, even if it means using the various biometric modalities to achieve this particular goal.

But in stark comparison, especially here in the United States, we take our standard of living for granted, thus we can afford to have a "nay say" attitude toward those particular technologies that would give us a better life, even if it means using biometric technology.

In other words, we know that our standard of living will continue to be the same irregardless of the choices we make to securing both our professional and personal environments and ways of living. Social science scholars often point to this flaw and weakness in our thinking as the American Society, as to why we always tend to be the prime targets of cyber-based attacks and threats.

So as one can see, changing the overall mindset of the American Public (as well as those of the other developed nations around the world) into adopting and using biometric technology will take an entire generation, if not a lifetime, to conquer. In other words, do not expect the adoption rate of biometric technology to pick up all at once here in the United States; rather, it will take many little steps to achieve this gargantuan purpose and objective. But in the end, perhaps very small steps toward a positive rate of adoption are far superior to no steps at all being taken.

Probably one of the best examples of this is right after the tragic and horrible incidents of 9/11. Just after these terrorist attacks, the use of biometric technology was acclaimed by the American Press to be the ultimate security tool of all, which would thwart off any future potential terrorist attacks. The biometric modality that got the most attention in this regard was that of facial recognition.

In fact, because of all of this media attention, the stock market price of many of the publicly traded biometric companies also skyrocketed as well. It almost felt like the Internet Boom once again, which transpired in the late 1990s. But after a brief period of time, after facial recognition failed to live up to all of the hype and expectations that were laid down before it, the American Press totally decried the use of it and made claims that it was a technology that would never prove its worth in securing and fortifying the American borders.

As a result of this, the American Public's already poor perception of biometric technology was only fueled further by these negative stories and reports by the American Press and media. But truth be told, facial recognition never failed to live up to its expectations in the technical and scientific sense.

For instance, at this particular point in time, facial recognition was still a biometric technology still in its infancy, which was totally ignored by the American Press, and as a result, went totally unnoticed by the American Public. Facial recognition merely failed to live up to the test and scrutinization of the American media. The following reasons can be cited:

- The issue of privacy rights:
  Once again, this concern or problem is a recurring one that occurs almost on a daily basis with biometric technology. In our previous book, the concept of privacy rights was more critically explored, and a specific definition was provided for it as well. It was further discussed that there are three main components that are attached to privacy rights, and they are as follows:
  - Anonymity;
  - Tracking/Surveillance;
  - Profiling.

  The first one deals more with the civil liberties groups, the second takes issue with the Federal Government (the fear of "Big Brother" watching), and the last one deals primarily with law enforcement. In reality, though, it is this last issue that the American Public takes the most fear of biometrics with, especially with the recent news stories that law enforcement across all levels (Federal, State, and Local) are receiving. This fear of law enforcement officials misusing the use of the various biometric modalities also transcends down to various age groups, especially those of senior citizens, the younger age group, and even especially the religious groups in the United States making claims of the "Mark Of The Beast". It should be noted that the US Constitution does offer indirect provisions for the protection of privacy rights amongst US citizens as well as foreign nationals. According to social science scholars, these specific provisions have been one of the main trigger points for the controversy of privacy rights and civil liberties violations surrounding the use of biometric technology here in the United States.

- The fear of ID Theft:
  There is no question that ID Theft is one of the most prevalent forms of cyberattacks and threats today. This happens when an important piece of information and data about a particular individual is stolen in a very covert fashion. As a result, the ID Theft victim does not actually realize what has happened to him or her until literally years later. By then, the damage has already been done, and it is then often too late for the victim to rebuild his or her own identity once again. In many of these cases, it is very often the credit card number that is hijacked or intercepted and used to either make large-scale purchases or even use it as a vehicle in which to assume the total identity of the victim. A brief history of ID Theft was reviewed, as well as how the major attacks occur today in the digital era. This fear of ID Theft has also transcended into the biometrics industry as well. This fear is triggered in many citizens of the United States because one's physiological or behavioral self cannot be changed if the associated biometric template has been hijacked. This stands in stark contrast to that of the username/password or credit card number. For example, if any bits of these types of data or information were to be intercepted, hijacked, or even stolen, they can be easily reset and issued once again. But however, this is not the case with regards to biometric templates. If they are ever hijacked or intercepted, there is nothing that can be done about it. As it was reviewed in this specific section, a biometric template is nothing but just a mathematical file. For instance, a fingerprint recognition-based biometric template is represented as a binary mathematical file – just a series of zeroes and ones (for example, 1010100001111101010). If this mathematical file were to be hijacked, there is nothing that a hacker could do with it. For instance, it is not the same as having your credit card number stolen and being used to make massive amounts of purchases. To date, there have been no actual cases of ID Theft occurring from a biometric template being hijacked or covertly intercepted. There have been cases where latent fingerprint images have been left behind on an optical-based fingerprint recognition sensor. Scientific studies have shown that these latent images can actually be used to reconstruct the actual raw image of the fingerprint, from which the unique features can be extracted and new biometric templates created.

- The overall perception of biometric technology:
  With this issue, it is not the social belief of what biometric technology is that is impacted, but rather it is what the American Public thinks about biometric technology if they were to see an actual biometric modality or device. For example, if a particular individual were to see for the first time a fingerprint recognition device, or an iris recognition device, or even a facial recognition device, what will his or her initial perceptions about it be? There is no doubt that there will be some wonderment about it, but for the most part, there will be a lot of apprehension and perhaps even grave fear about having to use it. As this section explained, the American Public simply does not understand what biometric technology is all about or what it can do. In large part, it is the biometrics industry that is to blame for this, for their lack of taking the much-needed initiative to launch training and education programs for the American Public to truly understand the concepts of biometric technology. Now, this is not to say that the average American citizen has to understand in great technical detail as to how a facial recognition system actually works, but there should be enough knowledge prevalent so that the average American citizen can have an appreciation for the levels of enhanced security that the various biometric modalities bring to the marketplace, as well as for the benefits and the strategic advantages they bring in as well. Rather, instead, the American Public views the differing biometric modalities as a "James Bond" type of technology, or simply put, the "Garbage In Garbage Out" phenomenon. To a certain degree, this is understandable, because once we submit our physiological or behavioral samples to a particular system, we have no control as to how it will be processed or even stored. The latter is the major concern because many business entities (even our US Federal Government) do not disclose how the biometric templates are stored or what security measures are afforded to make sure that they are not the prime target for a cyber-based attack or threat. In order to help calm these fears and anxieties amongst the American Public, the biometric vendors here in the United States have now started to make great efforts to make the various biometric modalities available today to possess a strong level of ease of use for the end user. Another factor that lends itself to the very poor perception of biometric technology here in the United States is that the processes that are used to develop the biometric devices amongst each and every different Biometric Vendor are proprietary, especially in the way of mathematical algorithms that are used. For example, the biometric vendors have their own mathematical algorithms, which are used to create both the enrollment and verification templates as well as to extract the unique features from the raw image that is captured. But

this tends to be a gray area because these algorithms tend to be the main revenue generator (or literally the "bread and butter") for these biometric vendors. But on the same token, this lack of transparency has also led to a lack of standards and best practices within the biometrics industry. If there was a list that could be easily accessed by the American Public, perhaps in the end biometric technology would not be viewed as a so-called Blackmagic technology.

- The ergonomic factor in the biometric modalities:
The last section just reviewed now lends itself to this new section. When the first biometric devices came out (which were actually the hand geometry scanner and the fingerprint scanner), they were very bulky to deploy and implement.

  An example of hand geometry recognition is illustrated below:

  Isometric Biometric Identification Hand Geometry Recognition Stock Vector (Royalty Free) 2141792595 | Shutterstock

  As a result of this, the American Public was very hesitant to use them because of the way these particular modalities literally looked. This is even true for the first retinal recognition device. This too was very bulky as well as very intimidating to use for the American Public, because the end user literally had to place their eye into a receptacle which was very cumbersome and tedious to use as well. This totally deteriorated any further use of the retinal recognition devices, and today, while it is a very powerful biometric modality to use for physical access entry market applications, in reality, it is hardly deployed and implemented anymore. Because of this, the biometrics industry here in the United States is now starting to realize that the way a particular biometric device looks to an end user (from an engineering standpoint) is very important in how it will be subsequently adopted and used later by the specific end user population. Therefore, there is now a movement from within the biometrics industry here in the United States to make the various biometric modalities very easy and, most importantly, comfortable for the end user to have their identity either verified or confirmed. Also in this section, a scientific study was also examined, which further stressed the importance of ergonomic design in the various biometric modalities in order to increase the adoption and acceptance rate of biometrics here in the United States. Also, this particular scientific study also stressed that the biometric device must not look intimidating either in the eyes of the American Public.

- The phenomenon of function creep:
As this section describes, biometric technology can be used either in a standalone mode (this is the instance where one biometric device is enough to conduct all of the verification transactions of the end user population-this is primarily used by the very small- to medium-sized

businesses in Corporate America), or the various biometric modalities can be used in conjunction with another (such as a fingerprint recognition device being used with a facial recognition device and even a third biometric modality such as iris recognition), or even be used with other non-biometric security technologies (such as a fingerprint recognition device or even an iris recognition device being used with a router or a firewall as a means of logical access entry). These specific instances of layered security solutions (in other words, biometric device with a biometric device, or a biometric device with a non-biometric device) are also known technically as "Multi-Modal Security Solutions". As a result of this, there is a lot of information and data interchange between all of these security devices, and at times it could occur that the information and data collected from one security device could accidentally be picked by another security device and used in a non-intentional format. Or, it could also be the case that the biometric templates in one database could be accidentally transferred into another database containing information and data from another security device. This phenomenon is known as "Spillover", or in more technical terms, it is also known as "Function Creep". Although this phenomenon occurs primarily internally within businesses and organizations across Corporate America, it has actually become quite transparent to the American Public. The primary reason for this is that the US Federal Government is by far the largest customer of biometric solutions from the biometrics industry and is also the biggest awarder of contracts (in terms of biometric technology procurement and deployment) to the private sector. It should also be noted as well that the US Federal Government also maintains some of the world's largest biometric databases, which are primarily used by law enforcement agencies not only here in the United States but worldwide as well. Because of this, there is great fear and angst amongst the American Public as to how the biometric templates which are stored in these respective databases will be used against them. For example, will the biometric templates stored in one database within a different agency of the US Federal Government be easily accessed by a local law enforcement official if a citizen is pulled over? In other words, the American Public wants assurances that the information and data that are stored about them will not be used covertly against them by the US Federal Government in other areas of their lives due to this "Function Creep" phenomenon.

- The legal considerations of biometric technology and its respective uses:
  Without a doubt, technology, no matter what it is, what purposes it serves, or even the specific market segments it delivers solutions to,

evolves and moves at a very quick and very rapid pace. Probably one of the best examples is that of wireless technology, especially that of the smartphone. This is also true of biometric technology. It is also advancing at a very rapid pace to the point now that it is very miniature, and the cost has become so low now that even the simplest biometric device can now be acquired at a local office supply store. But the legal framework that surrounds the use and deployment of biometric technology, however, moves quite literally at a snail's pace. For example, if a fifth-generation fingerprint recognition device were to come out, the courts here in the United States are still working on crafting the legal framework for the first version of this specific fingerprint recognition device. This very slow process of creating legislation around biometric technology has also proven to be a major hindrance for the adoption of biometric technology here in the United States. The following are the specific legal ramifications surrounding the use of biometric technology:

- The legal status of a biometric template:
  In this regard, is a biometric template considered to be private or public data? Or should it be classified as "Anonymous Data"? In this regard, the European Union has made much greater progress than the US Judicial System in ascertaining and formulating these specific classification schemes as they relate to the biometric templates that are created by the various biometric modalities.
- The storage of the biometric templates:
  One of the biggest legal issues here is that who owns the biometric templates once they are actually stored into a database? For example, is it the individual person, or is it the entity that is responsible for maintaining the database, or is it the Federal Government that owns these specific biometric templates? The other major legal hurdle in this regard is who addresses the security issues for protecting these biometric-based databases. Is it once again the Federal Government, or is it the entity whom is responsible for it? Also, how much security is deemed to be enough? Are there any legal precedents that need to be established in this regard?
- The usage of smart card technology:
  As it was mentioned earlier in this chapter, smart cards possess a special memory bank that can hold the information and data about the card bearer, including the biometric templates of that particular individual. But the legal question that arises here is who actually owns these biometric templates. Is it once again the card bearer, or the smart card vendor who actually manufactured the smart card technology being used?

- Can biometric technology be made into a mandatory security mechanism:
  This particular legal framework lends itself to Corporate America. For example, what if a place of business or organization has deployed the use of biometric technology in order to confirm the identities of their respective employees? Can the management team at this place of business or organization force the employees against their own will by mandating to their employees that they have to use the biometric system in place? Or will the place of business or organization will be required to offer other means of verification and/or identification if the employee refuses to utilize the biometric system set in place?
- The legal admissibility of biometric templates in a court proceeding/trial:
  As we all know, it took quite a long time for DNA to be recognized as a piece of evidence which can be legally used in a legal case in order to win a guilty verdict or exonerate a convicted individual. The same holds true for biometric templates. At the present time, there is no legal precedent established if a biometric template can be used with absolute certainty in a criminal trial or proceeding. Rather, it is up to the individual judge presiding over these particular legal cases if a biometric template can be introduced as legal evidence or not. Because of this judgment call, the fear resides in the American Public and even amongst the legal community if there will be fairness to a criminal trial or proceeding if biometric templates can be introduced as evidence, much in the same as DNA being used as evidence.

In both of these gargantuan market applications, biometric technology is being used quite heavily as one of the predominant means of security; in other words, using the various biometric modalities to fully and unquestionably ascertain and conform the identity of the individual who is either traveling overseas (making use of the e-Passport) or is voting in an election (making use of an e-Voting mechanism in order to have a ballot cast and tallied).

Many countries around the world have now adopted both of these types and kinds of infrastructures just described, but once again, it is the developing nations of primarily Africa and Asia that have been at the forefront of both of these market segments.

For example, the citizens of many of these developing nations simply have not had the means at their possession in order to have the ability to visit a foreign country. But with the possession of an e-Passport and now the e-Passport infrastructure within their own country, the ability to travel for the citizens of these nations is now possible.

The same also holds true for that of the e-Voting infrastructure. Once again, in many of these developing nations, the citizens have never been officially recognized by their respective governments as unique and single individuals. As a result, these people could not have the ability to cast their ballot in the elections that were held in their respective countries.

Even if they had the chance to cast a ballot, it was done so using the traditional methods of using an old-style paper ballot. With these old methods, the government and election officials whom were monitoring a particular voting precinct could quite literally throw away a paper ballot, or be even further succumbed to bribery and financial temptation in order to purposefully sway the outcome of a particular election.

But with an e-Voting infrastructure that makes the use of biometric technology, the citizens of these developing nations are now afforded the liberty of not only being counted as individual and unique citizens in the eyes of their own respective governments, but now they also possess the right to vote, which is a right they never enjoyed before.

With the use of the biometric modalities in the e-Voting infrastructure, not only will an individual's vote and ballot be counted, but there is now an irrefutable audit trail that will have been created of this vote and stored permanently in the e-Voting infrastructure's various types and kinds of databases.

Thus, if there are any charges of corruption or voter fraud in any given election, there is now an irrefutable and electronic audit trail which can be queried for and pulled up immediately from the respective databases in order to resolve any disputes or claims of election fraud or corruption.

But once again, in stark contrast, it is the developed nations, and in particular that of the United States, which so far have not adopted either an e-Passport infrastructure or even an e-Voting infrastructure.

One of the primary reasons for this is that it is our own Constitution that guarantees the right to American citizens that they will be counted as unique individuals in the eyes of the US Federal Government, and that no matter what, we as the American Public will always possess the right to vote and that our ballot will be counted and tallied under any type or kind of negative circumstance that might exist or come out.

In this regard, perhaps Social Science scholars could even technically find fault with the US Constitution as being probably the biggest roadblock or impediment for the poor adoption rate of biometric technology here in the United States.

A previous book of ours reviewed quite a bit of detail about the development of the e-Passport infrastructure and how the use of the various biometric modalities is now coming into critical play within it. More specifically, this chapter reviewed the following topics:

- The origins and the history surrounding the evolution and usage of the traditional paper passport:
  This section looked at how the traditional paper passport evolved and all of the stages of growth it went through. This includes such discussions as to how it was transformed from the standards set forth by the International Civil Aviation Organization (also known as the "ICAO") to being a mechanism that could be read by a machine-based reader (this particular type of paper passport is also known as a Machine Readable Passport). Also, the different types and kinds of traditional passports that are available today were also examined, along with their key and distinct features.
- The transition from the traditional paper passport to the e-Passport infrastructure:
  This section of the chapter examined some of the weaknesses that are associated with an e-Passport infrastructure, in particular the security risks and threats that are posed to it. The basics of the e-Passport infrastructure (both from a technical and social perspective) were then reviewed, as well as the advantages and the disadvantages it also possesses. The legal ramifications of the deployment and implementation of an e-Passport infrastructure were also reviewed, both from the standpoint of the country of origin and the country of destination, as well as the financial considerations and burdens, which also must be taken into consideration by the government of a country that is considering the deployment and implementation of an e-Passport infrastructure from within their own borders.
- The mechanics and the engineering design behind the e-Passport:
  This particular section examined in very close detail the processes that are involved in actually creating an e-Passport-based infrastructure. The following topics were then further analyzed:
  - The software for the IT structure:
    Such ramifications as the software tools needed for the capture of the respective biometric information and data, performing the needed quality control checks, as well as the software needed to house the biometric databases, were probed into.
  - The hardware for the IT structure:
    Such concepts as the contactless Integrated Circuit (IC) smart card-based microchip and the e-Passport reader (which essentially reads the information and the data stored onto the memory banks in the e-Passport) were also looked at. The usage of a public key infrastructure (PKI) structure to help secure and fortify an e-Passport infrastructure as well as the issues of the storage and processing of the biometric templates (and their associated pieces of information and data) and the placement

of the microchip within a particular e-Passport as well as a scientific study which examined the effectiveness of the radio frequency ID (RFID) protocol in an e-Passport infrastructure were also critically reviewed.

- The security vulnerabilities and weaknesses associated with the e-Passport:
  Just like the traditional paper passport, the e-Passport infrastructure also suffers from a set of security weaknesses and vulnerabilities, especially those that are cyber- or wireless-based. In particular, the following security threats and attacks that can be posed to an e-Passport infrastructure were examined:
  - Cloning attacks;
  - Eavesdropping attacks;
  - Adversary attacks;
  - Weak entropy of the basic access control keys;
  - Lack of access rules;
  - Once valid readers;
  - Denial of service attacks;
  - Leakage of biometric data;
  - A weak encryption key;
  - Man in the middle attacks;
  - No key revocation.

  Also, a traceability attack posed to a particular e-Passport infrastructure was also examined from the standpoint of a scientific study.
- The social impacts of the e-Passport infrastructure:
  Since biometric technology is now being heavily employed in the various e-Passport infrastructures around the world, it too is now prone to many and great social implications. Our previous book carefully examined the following social issues surrounding the deployment and implementation of a particular e-Passport infrastructure:
  - How the biometric templates (and their corresponding pieces of information and data) will be stored and used by the respective governments of both the country of origin and the country of destination;
  - The issues of privacy rights and civil liberties Violations from a macro sense, which includes all of the countries around the world that have adopted and deployed a particular e-Passport infrastructure;
  - The severe lack of understanding of the citizens around the world of the benefits and strategic advantages that an e-Passport infrastructure brings to the government of a particular country;
  - Once again, the issue of the storage of the biometric templates not only in the e-Passport itself, but also how these biometric

templates will be stored as well as used in both the country of origin and the country of destination;

- The ergonomic design considerations of the biometric modalities that need to be taken into account by the government in the country of origin that is issuing the e-Passport to a particular citizen;
- The health and safety considerations when a citizen has to visit the e-Passport office in their country in order to have their biometric templates processed so that it can be placed not only into the microchip of the e-Passport itself, but also that of the Biometric databases which located both in the country of origin and the country of destination;
- The varying and differing cultural and social beliefs which are held amongst the citizens around the world regarding the usage of a particular e-Passport infrastructure;
- Specific strategies and recommendations were also reviewed to ascertain what the governments around the world can do to help increase the acceptance rate of the usage of e-Passports and their associated infrastructures amongst their own particular citizens.

Finally, at the end of our previous book, an actual case study was reviewed, which examined the procurement, deployment, and implementation of an e-Passport infrastructure in Senegal.

Both the e-Passport infrastructure and the e-Voting Infrastructure can be viewed as symbols of establishment and embodiment for those countries which deploy and implement such types of technologies.

For example, with the e-Passport infrastructure, to many nations this represents a certain way of life and prestige in which their citizens can travel abroad and bring their experiences back with them and perhaps use that to launch a business or an organization to help stimulate their respective economies.

With the e-Voting Infrastructure, this gives nations around the world, especially the developing ones, a sense of the image that their election processes have now become fair and free of corruption and bribery and that it is also transparent to the rest of the world.

For instance, once again, in many of the developing nations, the governments have been heavily criticized by the governments of the developed nations that far too often, their particular elections are rigged and are very much succumbed to bribery and corruption.

But as mentioned previously, with an e-Voting infrastructure in place making the use of the various biometric modalities that are available today, there is now not only fairness in these types of elections, but each and every citizen is now counted in the eyes of their respective governments and is now finally given an equal opportunity to cast their ballot for the respective candidate(s) of their choice.

And if there are any questions of unfairness in the election processes of these developing nations, as also discussed in this chapter, there is now an irrefutable audit trail that can resolve any discrepancies very quickly.

But quite interestingly again, just like the e-Passport infrastructure, the United States has still not yet totally adopted an e-Voting Infrastructure, for the reasons cited primarily before.

According to the social science scholars, this is quite a contradiction when compared to the rest of the developing nations that have adopted, deployed, and implemented an e-Voting infrastructure and are, in fact, using it quite effectively and efficiently, as well as successfully.

Our previous book also examined as well in extremely close detail to the e-Voting Infrastructure. The following topics were covered and analyzed:

- The history of voting:
  In this section of the chapter, a critical look at how the traditional methods of voting evolved was carefully examined. For example, the history of the voting process was looked at in a couple of countries, but the main thrust was placed upon the United States. The pitfalls and the setbacks in which voters faced were probed into, such as the use of property and religion either allowing or denying an individual the fundamental right to vote, as well as the passage of the various constitutional amendments that were required in order to bring out equality and fairness to all citizens in the United States. Also, a timeline was provided that characterized the major social movements in the history of the US voting system since its evolution.
- The traditional methods of voting:
  This portion of the chapter examined in very close detail the inception as well as the evolution of the traditional voting mechanisms that have been utilized in the history of voting in the United States, dating all the way back to the early 1700s. Throughout this unique and rich growth, five primary stages of the traditional methods also evolved and were covered in detail as well. It should be noted here as well that the first voting methods were based upon the "open outcry" system, and that by the late 1800s, the first true mechanical voting machines actually came to fruition. The first of these devices were first invented in the United Kingdom, then made their way to the United States. After the mechanical voting machines were eradicated in the late 1990s, the paper ballot method of voting came about, then the punch card method of voting came, and finally, the optical scanning ballot system transpired.

- The problems with the traditional methods of voting:
  There is no doubt that the traditional methods of voting had (and even continue to have today) their fair set of security vulnerabilities and threats. This part of the chapter does not examine this particular component, but it also took a unique perspective as well in order to determine why the voter turnout in recent elections has been so low (and which is also part of the reason for the very slow adoption rates of an e-Voting infrastructure being deployed here in the United States). The following are cited reasons for this:
  - The overall dissatisfaction with the candidates in time of election;
  - The current economic conditions in the United States at the time of a particular election;
  - Electoral competitiveness;
  - The type and kind of election that is taking place;
  - The age of the voters in question;
  - The race and ethnicity of the voters;
  - The gender of the voter;
  - Hereditary factors;
  - Voter fatigue;
  - Cultural factors;
  - Institutional factors;
  - The level of saliency;
  - The ease of voting.
- The transition to e-Voting:
  The primary catalyst for this was the great closeness of the 2000 Presidential Election. A formal definition of the e-Voting infrastructure was provided, as well as the major advantages and disadvantages of an e-Voting infrastructure if it were to be implemented here in the United States. The first type of e-Voting infrastructure to be experimented with here in the United States was that of the direct recording electronic devices, and its major technical and electronic components were reviewed as well in detail, which are as follows:
  - Electronic voter lists;
  - Poll worker interfaces;
  - Interfaces for voting;
  - Special interfaces for disabled voters;
  - Interfaces for displaying the output/results of an election;
  - Results transmission systems;
  - A more modern means of displaying election results;
  - Confirmation code systems.

  The major technological classifications of an e-Voting infrastructure were also examined, and they are as follows:
  - The direct recording electronic voting machine;

- The optical mark recognition (OMR) system;
- The electronic ballot printer system;
- The Internet voting system.

It should be noted here as well that a particular e-Voting infrastructure can be deployed and implemented in either a controlled environment (where it is being used at a polling station) or in an uncontrolled environmental setting (this is typically where Internet-based voting is being utilized). The differences between these two environments as well as their respective advantages and disadvantages were also critically reviewed, as well as the use of open source software versus the use of closed in creating the electronic-based modules in an e-Voting infrastructure.

- The security vulnerabilities of e-Voting:
  Just like its traditional voting counterpart, an e-Voting infrastructure also possesses its set of security vulnerabilities, as was examined in this section. These vulnerabilities include the following:
  - Voter fraud;
  - Improper security measures on the boxes containing the votes;
  - Election corruption;
  - Voter identity authentication;
  - Adware, malware, and spyware;
  - "Buggy" software;
  - Distributed denial of service (DDoS) attacks;
  - Insider attacks;
  - Spoofing;
  - Man in the middle attacks.

  Also, a specific scientific study was reviewed as well, which closely examined the security threats posed to an e-Voting infrastructure, especially from the client side (this is when Internet-based voting is used and the voter needs to access a special type or kind of graphical user interface from either their computer or wireless device in order to cast and transmit their ballot).

  In an effort to combat these major security threats and risks that are posed to an e-Voting infrastructure, making use of the principles of cryptography was also examined in this section. However, as it applies to the e-Voting processes, a cryptographic-based infrastructure must also meet a set of stringent requirements, which are as follows:
  - It must ensure privacy;
  - It must preserve the authentication process;
  - It must ensure the level of accuracy of the e-Voting infrastructure;
  - The maintaining of the secrecy of the e-Voting process;
  - Non-coercion by third parties;
  - The verifiability of the e-Voting process must be established.

A specific type of Application Layer Cryptographic Protocol was examined as well, which can be used in an e-Voting infrastructure; this is known as the "Pynx" software package, and it includes the following components:

- The key principle;
- The service protection;
- The security techniques;
- Restricted computational processing demands;
- The use of audit trails;
- The principle security requirements are met or exceeded;
- Cryptographic hashing functions;
- Encryption keys;
- Digital signatures.

With these principles in mind, the possibilities of using a PKI in an e-Voting process were also examined. This would make use of public keys and private keys, the advanced encryption standard, as well as the digital signature algorithm.

- The use of biometrics with e-Voting:

  This component of the chapter looked closely at how the various types and kinds of biometric modalities (both from the physical biometric and behavioral biometric standpoints) can be incorporated and utilized in an e-Voting infrastructure. These modalities can be used from three standpoints:

  - Physical access entry;
  - Time and attendance;
  - Single sign-on solutions.

  As it was also reviewed, the most common biometric modalities that are being used in e-Voting infrastructures worldwide include fingerprint recognition, hand geometry recognition, iris recognition, and even facial recognition. The various policies and directives being used to create a list of best practices and standards in the creation, deployment, and implementation of an e-Voting infrastructure were also carefully examined. These include the following:

  - The ISO/IEC 7816-11;
  - The Common Biometric Exchange Format Framework;
  - The XML Common Biometric Format;
  - The ANSI B10.8;
  - The ANSI/NIST ITL 1-2000;
  - The ESIGN-K;
  - The DIN V64400;
  - The BioAPI;
  - The Biometric Application Protocol Interface (API).

  Launching an entire e-Voting infrastructure can be quite a project management endeavor, and many key variables need to be taken into consideration. Some of these specific variables include the following:

- The cost of the biometric modality;
- The metrics of the biometric modalities;
- The false reject rate (also known as the FRR);
- The false accept rate (also known as the FAR);
- The resiliency of the biometric system from ID Theft attacks;
- The fail safety component of the biometric system.

The use of the principles of biocryptography in an e-Voting infrastructure was also examined, and a scientific study that further explored this use was also reviewed, which also implemented the usage of a virtual private network, or VPN. Also, an e-Voting Infrastructure can make use of either an unimodal biometric system (where just one biometric modality is used, such as fingerprint recognition) or a multimodal biometric system (where two or more biometric modalities are used, such as fingerprint recognition and iris recognition being used sequentially together). Another scientific study was also examined in this regard, which looked at the security vulnerabilities of using just a unimodal biometric system in an e-Voting infrastructure. A third scientific study was also reviewed as well, which examined the benefits and strategic advantages of using a multimodal biometric system in an e-Voting Infrastructure. It appears that the use of an e-Voting infrastructure in an uncontrolled environment (such as that of remote, Internet-based voting) will be the trend of the future. Therefore, a cloud-based infrastructure could be used quite easily and effectively here as well (also known as "Biometrics as a Service", or "BaaS" for short). In this regard, another scientific study was also closely looked at, which examined how an e-Voting infrastructure would actually take place in a virtualized or cloud-based environment.

Our previous book also reviewed the deployment and implementation of an e-Passport infrastructure in the country of Mozambique.

As it was discussed in this chapter, it is the e-Passport infrastructure as well as the e-Voting infrastructures that are being utilized the most around the world, especially in the developing nations such as those in Africa and Asia. But in the developed nations, such as that of Europe and the United States, the evolvement and growth of these market applications of biometric technology simply has not taken off yet where the rest of the world is at.

Numerous reasons and explanations have been cited for this, one of them primarily being the rights and the liberties that are guaranteed to American citizens by the US Constitution. In fact, it is not just in the areas of the e-Voting Infrastructure and the e-Passport infrastructures that have been lacking here in the United States, but there are also other market applications that utilize the various biometric modalities that simply have not had a quick adoption rate as well.

But as it was also mentioned at the beginning of this chapter as well, when compared to the other security technologies, it is biometric technology that leaves both a lasting technical and social implication upon the American society at large. The social science issue of how to increase the adoption rate of biometric technology here in the United States is a very complex one.

For example, not only are the impacts of privacy rights and civil liberties, which are guaranteed to us by the United Constitution, not only an impediment toward trying to escalate this particular adoption rate, but also, what makes this issue such a gray area is that it contains both macro and micro issues as well. For example, in terms of the former, it is the US Federal Government that is still by far the largest consumer of biometric technology.

The impacts of this can be felt across all industries here in the United States. For instance, whenever we travel, we are now subject to extra means of security at the major international airports, one of them being the so-called Body Scan. This technology can literally see through the inside of a person, even in the most extreme of the private regions.

Because of this, the American Public has claimed that the use of this Body Scanning Technology is an utter and sheer violation of privacy rights as well as civil liberties. Because of the negative impacts that were created by the usage of the body scanning technology, this has also had a negative ripple effect upon the overall acceptance of biometric technology by the American Public.

An example of a micro issue is that now the US Federal Government has started to introduce and mandate the usage of biometric technology for school lunch programs for the children of the families who are living well below the poverty line. With this, a student does not have to pay for their school lunch with cash; rather, they can pay for it by registering and enrolling their respective fingerprints into a biometric database.

Therefore, once they enter the line to procure their lunch for the day, all the student has to do is merely have their identity confirmed by the fingerprint recognition system, and from there, all of the food and the other consumable items are then paid for by the Federal Government. But despite these best intentions, many parents have totally and outrightly objected to the use of biometric technology (in this case, fingerprint recognition) to register their children into these subsidized school lunch meal programs.

Our previous book not only summarized the previous content that was covered in the preceding four chapters, but it also examined in a little bit more detail some of these other social and technical issues as well. For example, it was also pointed out in this concluding chapter as well that the type of biometric modality that is being used can also have a great impact upon the overall adoption rate of biometric technology here in the United States.

For example, in this regard, the oldest of the biometric modalities, such as that of hand geometry recognition and fingerprint recognition, will probably be the most widely accepted for usage by the American Public. This is primarily so because these two types of biometric modalities have been around the longest, and the ergonomic designs of these two particular biometric modalities have also become much more appealing and easy to use, thus helping to increase their adoption rate in the various market segments that they both serve.

But on the contrary side of this, the biometric modalities of facial recognition and retinal recognition would have a very poor and dismal acceptance rate with the American Public. In terms of the former, the nature of this particular biometric technology has always had a negative social connotation attached to it (primarily that of "Big Brother" watching over the shoulders of every day average American citizens). In terms of the latter, this particular biometric modality is just way too user invasive in order to be widely accepted by the American Public.

Interestingly enough, a subset of retinal recognition is that of iris recognition. As a sharp contrast, this particular biometric modality is actually starting now to receive a higher acceptance rate amongst the American Public. At one point in time in the last decade, iris recognition was also deemed to be just as user invasive as retinal recognition continues to be today.

But over time, iris recognition technology became much more user-friendly and "sleek" in terms of its particular ergonomic design, as well as in terms of its strong level of ease of use. As a result of all of these positive factors, the American Public has been much more receptive and accepting toward both the adoption and usage of iris recognition.

In terms of the behavioral biometrics, such as that of keystroke recognition and signature recognition, it is also highly expected that over the course of time in the future, they too will have a strong adoption rate here in the United States.

The primary reason for this is that these two particular biometric modalities are not only quite easy to use for the American Public, but they can also be very easily installed into an existing security-based or IT infrastructure quite easily and quickly as well.

But the reason why these two behavioral-based biometric modalities will need time to be accepted by American citizens is that when compared to the other physical-based biometric modalities, keystroke recognition and signature recognition are the least deployed, used, and implemented. Our previous book also examined three potential biometric modalities, which are currently in the research and development phases. They are as follows:

- Earlobe recognition;
- Gait recognition;
- DNA recognition.

So far, it is gait recognition that holds the most promise of being a valid biometric technology. With this modality, the identity of a particular individual can be confirmed by the way they walk or their unique strides, which he or she possesses. After gait recognition, the next potential biometric modality that holds great promise is that of DNA recognition.

A particular individual can have their identity confirmed just based solely upon the four unique strands they possess in their DNA (which are adenine, guanine, cytosine, and thyamine). But if DNA Recognition ever does become a viable biometric technology, it too will be prone to severe claims of privacy rights as well as civil liberties violations. The primary reason for this is that the usage of DNA is viewed by the American Public to be amongst the most private in terms of a physiological or biological snapshot being captured of our individual selves.

But however, despite the very negative images that are portrayed here in the United States, there are several growth areas of the usage of biometric technology, and these particular market segments could be those specific catalysts and triggers that are needed in order to spur up the adoption rate of biometric technology here in the United States.

These specific market segments include the following:

- Single sign-on solutions;
- Time and attendance;
- Mobility and wireless devices.

In terms of the first one, the American Public knows all too well of the security vulnerabilities of which the username/password combination possesses. It can be easily hacked into and intercepted as well by a malicious third party. Because of this, Corporate America has now literally mandated the use of much more stringent passwords, which are now often too long and complex to remember.

Because of this, many employees now write their passwords down on a post it note and literally attach it to their workstation monitor. This has become known as the "Post It Syndrome" and, of course, defeats all purposes of having a longer and much more complex password to remember.

This phenomenon has also led to an increase in the number of password resets that take place, and this can be an expensive proposition for any type or kind of business or organization. For example, on average, it costs at least $300 per employee per year for each password reset that takes place.

As a result of all of this, the use of biometric technology has been called upon. In this particular market segment, it is fingerprint recognition that is the most widely used biometric modality. The optical sensor can be attached within the actual workstation or be connected to it via a USB cable. With one swipe of the finger, the employee of the business or organization can be logged into their workstation in just a matter of minutes, hence its name, the "Single Sign On Solution". It totally does away with and eliminates the need for or usage of the username/password combination.

In terms of the second market application, many business entities here in the United States still utilize the old-fashioned method of the time card system in order to keep track of employee working hours and also for computing payroll as well.

But as one can see, this old-fashioned method poses a lot of problems and security threats as well. For example, it can take a lot of administrative manpower as well as time to enter all of the clock-in and clock-out times onto a separate spreadsheet. Also, the phenomenon known as "Buddy Punching" is also prevalent when using the old fashioned system of the timesheet.

This occurs when an employee calls in sick but still has his or her friend at the place of business or organization covertly and secretly punch in and punch out their time for them. The end result is that the employee who fictitiously called in sick will still get paid for the hours in which no levels of productivity occurred what so ever.

To combat all of this, once again, the use of biometric technology has been called upon as well. In this regard, it is hand geometry recognition and fingerprint recognition that are used the most, with iris recognition coming in a close third. With the usage of these various biometric modalities, there is now irrefutable proof and evidence now available as to the actual employees whom have worked a particular day, as well as their respective punch in and punch out times.

Not only is the "Buddy Punching" security risk totally eradicated, but so is the need to hire extra administrative personnel to record the punch-in and punch-out times onto a spreadsheet and to compute payroll as well. All of this is done automatically and quickly with the biometric modalities just mentioned.

With regards to the last market application, the use of the smartphone has not exploded just worldwide, but it too has grown and advanced quite rapidly here in the United States as well. In reality, the smartphone has become literally an extension of both our personal and professional lives. We use to communicate not only by phone, but to send text messages and various E-Mail messages as well. We conduct e-commerce transactions on it and even use it as a calendar to help us keep track of our daily agendas, activities, and meetings.

But there are two inherent security weaknesses with the smartphone: They can either be lost or easily stolen. In any case, it is quite likely that it will fall into the hands of a malicious third party. From here, a kind of information and data can be quite easily hijacked, and from here, it can become the launching pad to launch covert ID Theft attacks from.

To help secure their smartphones, the major wireless carriers have long mandated the use of a password or a unique PIN number to help secure them. But as it has been illustrated, the use of the password has now become the weakest chain in the link, especially when it comes to securing smartphones.

Therefore, once again, the use of biometric technology has been called upon to help secure smartphones as well as other types and kinds of wireless technologies. In this fashion, fingerprint recognition has been the most widely used biometric modality, as a tiny optical sensor can be placed onto the hardware of the smartphone. The pioneer in this area has been the Apple Corporation, with its specialized "TouchID" fingerprint recognition technology.

Iris recognition and facial recognition have also been used, and in this regard, it is the camera of the smartphone that can be used to capture the image of the iris or the face, respectively. From this point, specialized mobile apps can then be downloaded onto the smartphone, which can then take these iris- or facial-based raw images, extract the unique features, and subsequently create the required enrollment and verification templates.

The adoption rate of biometric technology here in the United States will take a very long time to occur. But there is no doubt that one day it will actually happen, given not only the need for more sophisticated versions of security technology but also to help combat the ever-growing and constant threats of cyber- and Internet-based attacks and threats.

However, there is yet another critical, third piece that needs to be examined when it comes to biometrics. That is the software development and application side of it.

For example, whenever we see an actual biometric device or view an image of it, we often think that it is all about hardware-related aspects only. Yes, the hardware component (which is the actual biometric device) is a critical piece; after all, this is where the biometric templates can be stored, and the respective verification and/or identification transactions actually take place.

But the truth is, many biometric devices can all be interlinked together and even be connected to a central server.

At this level, the database containing the biometric templates and the verification and/or identification transactions can be stored here. Therefore, the need to create customized software applications and interfaces becomes important. At the present time, many biometric vendors offer their own "canned" suite of software applications in which an end user can easily install and use literally straight out of the box.

But many different businesses and organizations have their own unique set of requirements when it comes to the software application side of a biometric system implementation. As a result of this, many of the biometric vendors also offer (along with their "canned" suite of software applications) what are known as "Software Development Kits", or also known as "SDKs" for short.

With this, the IT development team at a business or organization can use these different and various SDKs to build their own customized applications.

But however, the software applications that are developed as a result of using these SDKs are often not known to the public, or for that matter even in the IT Security and biometrics sectors. In the software development world, there is often a need for having a "Best Practices" list in order to create the most robust and efficient software code possible, and these particular lists do often exist.

But however, in the biometrics industry, when it comes to the development of software-related applications, no such "Best Practices" list exists.

This stems from the lack of knowledge of the actual software applications that are being coded and developed.

This book is an update to our very first book on biometrics, which is entitled: "Biometric Technology: Authentication, BioCryptography, and Cloud Based Architecture". The updates in this book will include the following:

- A deeper dive into the AWS and Azure as it relates to the concept of BaaS.
- Other applications of biometric technology that include single sign solutions and multimodal solutions.
- A discussion of the use of APIs in biometric technology and the security protocols that need to be taken with them.
- A review of generative AI.

## REVIEW

This chapter provided an overview of the science and nature of biometrics. The goal is to provide a broad underpinning as to what biometrics is all about. Its primary goal is to give the background needed to comprehend the rest of the topics covered in this book.

In summary, this chapter provided an introduction of what unique features are all about; an exact definition was provided of what biometrics is all about (as well as what it is **NOT**); a summary of the biometric technologies that are available (physical biometrics versus behavioral biometrics) was reviewed; a specific definition of recognition was examined as well as its components (verification, identification, authorization, and authentication); what a biometric template is (as well as the various types and kinds of biometric templates that are used today, especially those of enrollment and verification templates); the most important biometric technology and standards were explored in great detail; an overview of sensor technology and how it is used in biometrics was closely looked at; as well as a brief survey of the major biometric applications (such as physical access entry, time and attendance, logical access control, surveillance, and law enforcement).

Chapter 2

# Physical and behavioral biometrics

## THE BIOMETRIC TECHNOLOGIES OF TODAY – PHYSICAL AND BEHAVIORAL

This chapter will now take, if you will, a piece of Chapter 1 and explore it in much more exhaustive detail. And that piece is the biometric technologies which are available today. Each biometric technology which was presented in Chapter 1 will now be further expanded upon and reviewed.

At this point, before we move onto our detailed discussion and analyses of the biometric technologies which are available on the marketplace today, it is very important to turn back to the topic of physical biometrics and behavioral biometrics. As it was mentioned in Chapter 1, physical biometrics involves taking a biological or physiological snapshot of a part of our body, and behavioral biometrics involves taking a picture of our unique mannerisms, which makes us different from everybody else.

But now, we need to provide much more precise, scientific definitions to these two terms. Physical biometrics can be defined as follows: "Acquiring physical biometric samples which involves taking a measurement from subjects. This does not require any specific action by the subject. A physical biometric is based primarily upon an anatomical or physiological characteristic rather than a learned behavior".[1]

Behavioral biometrics can be defined as follows: "Acquiring behavioral biometric samples requires subjects to be active. They must perform a specific activity in the presence of a sensor. A behavioral trait is learned and acquired over time rather than based upon biology.[2]

## THE DIFFERENCES BETWEEN PHYSICAL BIOMETRICS AND BEHAVIORAL BIOMETRICS

As one can see, based upon these two definitions, some very subtle differences between physical biometrics and behavioral biometrics can be observed. Probably the biggest difference to be noted is the amount of activity required by the end user. With physical biometrics, no active part is needed on part

DOI: 10.4324/9781003562146-2

of the individual in order to collect that individual's physiological and biological unique features (whether it be the hand, the finger, the iris or retina, or even the vein patterns present from underneath the palm or the fingerprint). Of course, the end user must be cooperative in order for an effective sample to be captured (this fact is true of all of the biometric technologies).

In other words, physical biometrics can be more or less considered as "garbage in, garbage out". The image is captured, unique features are extracted, and either the individual is verified or not verified. But with behavioral biometrics, a specific function, or an active part, must be carried out (such as typing on a keyboard or signing your name), which is learned over time.

As a result, because of this learned behavior, deviations, or changes in the behavioral-based biometric templates which are collected can drastically occur over time, at a level much more so than physical-based biometric templates. Hence, research and development is already underway in which a behavioral-based biometric system can literally learn and take into account such changes in the behavior and mannerisms which can occur over the lifetime of an individual.

A perfect example of this would be the use of neural network technology. With this type of technology, computer-based learning and reasoning takes place in an attempt to closely follow the actions and patterns of the human brain.

Another difference between physical biometrics and behavioral biometrics which is evident from the above definitions is the number of measurements taken in order to extract the unique features. Although in both types of technologies, multiple images or multiple samples are collected. But in physical biometrics, only one composite image is utilized to extract the unique features from. Thus, with physical biometrics, it is considered only one measurement is taken.

With behavioral biometrics, multiple samples are collected, but a composite sample is not created in which to extract the unique features from. Rather, statistical profiles are created from each individual sample in order to positively verify or identify an individual. For example, with keystroke recognition, an end user has to type multiple lines of text. Statistical profiles (using hidden Markov models (HMMs) primarily) are created on the basis of each line of text typed, and from there, the unique patterns are noted, and then the individual is either fully verified or not verified at all.

## WHICH ONE TO USE? PHYSICAL BIOMETRIC OR BEHAVIORAL BIOMETRIC?

A question which often gets asked at this point is what type or kind of biometric system should be used? In other words, should a physical-based or behavioral-based biometric system be utilized? As it was eluded in Chapter 1, and which will be covered in more detail in later chapters, a big factor upon which biometric system to choose will largely depend upon the system analysis and design study which should have been conducted prior to the procurement and acquisition of a biometric system.

And of course, another big factor is the type of security application upon which you are planning to use biometrics. For instance, as a C-Level Executive, are you looking at verification or identification type of scenario? Or in other words, are you going to confirm the identity of an individual on a one-to-one level, or on a one-to-many basis?

When it comes to verification scenarios, just about any biometric technology will work, whether it is physical or behavioral based. Probably the more robust biometric technologies to be used in this fashion would be that of fingerprint recognition, iris recognition, hand geometry recognition, and vein pattern recognition.

But, when it comes to identification scenarios, the only choice to use would be that of the physical-based biometrics. Behavioral-based biometrics are simply not robust enough to capture the identity of an individual in a large database. Also, as it was discussed, at this point in time, behavioral-based technologies simply do not possess the capability to capture the dynamic changes in the behaviors or mannerisms of an individual which occur over the course of time.

In this regard, once again, physical biometrics becomes the prime choice to be used for identification types of applications. Fingerprint recognition is the best choice here, as it has been used the most, and tested across very large databases.

Also, another critical factor in the choice of a physical or behavioral based is the ease of the acquisition of the respective biometric templates upon the specific environment they operate in. In other words, it is very important to choose a physical- or behavioral-based biometric which will fit into the environment you want it to operate in.

For instance, behavioral-based biometrics will work in small office-based environments, and physical-based biometrics such as hand geometry recognition is more suited for usage and extremely harsh environments, such as factories and warehouses. Facial recognition is well suited for covert types of applications, such as tracking individuals in airport settings, or large sporting venues. On the opposite side, none of the behavioral-based biometrics are suited for large-scale environments.

Apart from all of these factors just described, there are other hurdles included as well, when choosing either a behavioral or physical-based biometric system. These other factors include the following:

1. The environment in which the biometric system will operate in;
2. The network bandwidth and data transmission needs in the transfer of the biometric templates as they traverse across the network medium;

3. The size of the population and demographic specifics upon which the biometric system will be used in;
4. The environmental comfort of the biometric system for the end user (also known as ergonomics);
5. The ability of the biometric system to operate in conjunction with legacy security systems;
6. The end user acceptance level of the biometric system which will be deployed.

Remember in the end, there is no such thing as the perfect biometric system. They all have their strengths and weaknesses, as well as their technological flaws. Therefore, great thought needs to be given ahead of time to the biometric system which will best fit the needs of your business.

## FINGERPRINT RECOGNITION

As mentioned back in Chapter 1, fingerprint recognition has been the one biometric technology which has been around for the longest time. The use of fingerprints goes back all the way to the 1500s or so, and ever since then, it has been associated with law enforcement. In fact, the fingerprint has become the de facto standard used in law enforcement (Figure 2.1).

In fact, fingerprints were founded as far back as 2000 BC, as a means of a legal signature in the very primitive business transactions which took place during that time. The first true research paper which attempted to examine the unique structures of the fingerprint was published around 1694, and the very first fingerprint classification system evolved in 1823, by a scientist known as Jan Purkinje. And of course, at the turn of the century,

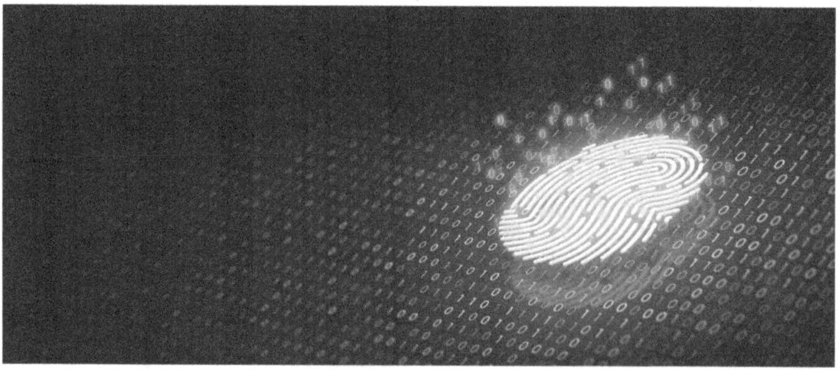

*Figure 2.1* This is an image of a fingerprint, showing the unique features that can be extracted. (Future Security Technology Fingerprint Scan Provides Stock Illustration 1941373864 | Shutterstock.)

beginning in the early 1900s, law enforcement agencies here in the United States started to use fingerprints as the primary means to track down known suspects and criminals.

The following statistics just further exemplify the popularity of using fingerprints as the primary means of the verification and/or identification of individuals:

1. By 1994, the FBI databases held 810,000 fingerprints;
2. By 2003, this number had swelled to more than 200 million fingerprints held in their databases;
3. By 2009, this number had escalated to 500 million plus fingerprints held in the FBI databases.

Obviously, in order to keep up with the growing demand for the use of fingerprints, the FBI devised the Automated Fingerprint Identification System, or AFIS, in order to automate fingerprint-based searches across levels of law enforcement, at the national, state, and local levels. And in order to keep up with the growing demands of this gargantuan database, the Integrated Automated Fingerprint Identification System, or IAFIS, was introduced, with enhanced features. The details of this can be seen in Chapter 1. An image of a fingerprint can be seen in Figure 2.1.

But, in the world of biometrics, the details of the fingerprint are broken down into three distinct levels:

1. Level 1: The pattern images which are present in the fingerprint;
2. Level 2: The minutiae points of the fingerprint (this is from where a bulk of the unique features are actually extracted from);
3. Level 3: This includes the shapes and the images of the ridges, and its associated pores.

## The unique features

It is important to note at this point that most of the biometric-based fingerprint systems only collect images at Levels 1 and 2 only; it is only the most powerful fingerprint recognition systems which collect Level 3 details and is used primarily for identification purposes. The Level 1 specific features include the following:

1. Arches: These are the ridges which just flow in one direction, without doubling back, or going backward. These only comprise about 5% of the features of the fingerprint;
2. Loops: In this feature, the ridges go backward and go from either the left to the right or the right to the left. There are two distinct types of loops: (a) Radial loops which go downward; and (b) the ulnar loop

which goes upward on the fingerprint. These make up 65% of the features within the fingerprint;
3. Whorls: The ridges in the fingerprint make a circle around a core, and these comprise 30% of the features in the fingerprint.

In addition to the above features which are collected by a fingerprint recognition system, the actual number of ridges and the way that these ridges are positioned (specifically their orientation) can also prove to be a very distinctive feature as well and can help to contribute to the verification and/or identification of an individual. Other distinctive features which can be extracted, but are not as commonly used, include the following:

1. Prints/islands: These are the very short ridges found on the fingerprint;
2. Lakes: These are the special indentations/depressions located right in the middle of the ridge;
3. Spurs: These are the actual crossovers from one ridge to another.

## The process of fingerprint recognition

Fingerprint recognition is one of those biometric technologies which works not only well for verification types of scenarios, but it also works very effectively for identity applications as well, which is best exemplified by the gargantuan databases administered by the FBI. But whether it is identification or verification which is being called for, fingerprint recognition follows a distinct methodology which can be broken down into the following steps:

1. Raw data acquisition: The actual, raw images of the fingerprint are acquired through the sensor technology which is being utilized (a detailed review of various sensors can be seen in Chapter 1). At this point, a quality check is also included. This means that the raw images which are collected are eventually examined by the biometric system to see if there is too much extraneous data in the fingerprint image, which could interfere in the acquisition of unique data. If there is too much of an obstruction found, the fingerprint device will automatically discard that particular image and prompt the end user to place their finger into the platen for another raw image of the fingerprint to be collected. If the raw images are accepted, they are subsequently sent over to the processing unit, which is located within the fingerprint recognition device;
2. With the raw images which have now been accepted by the system, the unique features are then extracted and then stored as the enrollment template. If fingerprint recognition is being used by a smartphone, a smart card is then utilized to store the actual enrollment

template and can even provide for some processing features for the smartphone;
3. Once the end user wishes to gain physical or logical access, he or she then has to place their finger onto the sensor of the fingerprint recognition system, so that the raw images and unique features can be extracted as described above, and this becomes the enrollment templates. The enrollment and verification templates are then compared to one another, to determine the degree of similarity/non-similarity with one another;
4. If the enrollment and verification templates are deemed to be close in similarity, the end user is then verified and/or identified and is then granted either physical or logical access to which they are seeking.

As you can imagine, as C-Level Executive, the quality control checks which are put into place for the fingerprint raw images are rather extensive. The common error in thinking is that it is the enrollment and the verification templates which are the most important in any type of biometric system. While this is true to a certain extent, keep in mind that these templates are actually constructed and created from the raw images themselves.

## Fingerprint recognition quality control checks

Thus, the need to collect high-quality, raw images the first time is of paramount importance. The following are the quality control checks for fingerprint recognition:

1. Resolution: This refers to the total number of dots per inch (DPI), or also known as the total pixels per inch (PPI). Most fingerprint algorithms require a DPI or PPI resolution of at least 250–300 DPI, and to meet the stringent requirements of the FBI, a minimum of 500 DPI is required;
2. Area: This is the actual size of the scanned and captured fingerprint image. The specifications set forth by the FBI mandate a minimum scanned fingerprint area of at least 1 square inch. Anything less than that will not produce a good quality, raw image;
3. Frames per second: This is the total number of raw images the fingerprint recognition device sends to the processing unit. Of course, the higher number of frames per second means a much greater tolerance for any unwanted movements of the fingerprint on the platen;
4. The number of fingerprint pixels: This refers to the total number of pixels in the scanned image;
5. The dynamic range: This refers to the possible ranges which are available for the encoding of each pixel value; the FBI mandates at least eight bits;

6. Geometric accuracy: These are the geometric differences between the enrollment and the verification templates, and this is calculated via the deviations from the X and Y axes on the fingerprint template;
7. The image quality: This is the variable which refers to the indentifying, unique features in the fingerprint, such as the ridgeline patterns and the various minutiae which are extracted.

## Methods of fingerprint collection

Fingerprint images can be collected in one of two methods (or even both, if the need be). They are as follows:

1. Offline scanning;
2. The use of live scan sensors.

Offline scanning methods have been in existence since fingerprints have been used in law enforcement. These traditional methods have involved using an inked impression of an individual's finger, and then placing that impression onto a piece of paper. Subsequently, cards were utilized instead of paper in order to store the inked fingerprint image. These were kept in file cabinets, and when a suspect was apprehended, their fingerprints were then compared to the inked fingerprint image stored on the card.

Over time, as technology improved, these two-dimensional images were then scanned into a digital format, translated into a 500 DPI, in order to meet the FBI requirements and specifications. But, with today's fingerprint recognition systems, a law enforcement officer in the field with a wireless biometric device can scan a suspect's fingerprint and then have that image automatically uploaded into a central server, from where it can then be converted into a proper fingerprint digital image.

As one can imagine, the use of these traditional methods definitely has its fair share of flaws. For instance, it takes great skill to get a good inked impression of a fingerprint, too little ink used means that the print area of the particular finger can be missed, and too much ink used could very well obscure any features which were attempted to gather. But the offline method does possess one great advantage over the live scan method. And that is, a "rolled" impression of the fingerprint can be gathered, which means that a much larger image of the fingerprint can be captured, versus the much smaller scan areas of the live scan sensors.

To this degree, the question often gets asked is how are the latent fingerprints captured? Latent fingerprints are those which get left behind at a crime scene and are collected later in the course of an investigation. Offline scanning methods are used to collect these latent fingerprint images at a later point in time, and oftentimes, chemical reagents and processes are used to collect the fingerprints.

With live scan sensors, two methods of collecting fingerprint images are used:

1. Touch
2. Sweep

With the touch method, direct contact is required of the finger onto the biometric sensor. But using this type of method possesses some grave disadvantages. For instance, the platen can become dirty or smudged quite easily, any fingerprints left behind on the sensor or platen by another user previously can lead to substantial errors, and in terms of dollars and cents, there is a positive correlation between the size of the scan area and the actual cost of the sensor. Also, some sensors may not be able to capture raw images if the finger is off more than by 20 degrees from the platen.

Unlike the touch method where only finger is scanned, the sweep method captures images of multiple fingerprints as they are read by the fingerprint sensor, and from that point, multiple images are captured at the end of the sweep. From these multiple images, one composite image is created. The quality of the raw images which are captured is also partially dependent (as well as the other variables just described) on the accuracy of the reconstruction algorithm whose primary function is to compile the multiple raw images into one image of the fingerprint.

The biggest disadvantage of the sweep method is that a much higher rate of error is introduced due to the variance in the sweeping rates and the angles. It is the sweep method which is primarily used by fingerprint recognition devices today, and there are three types of sensors which are available:

1. Optical sensors: These are the most commonly used;
2. Solid state sensors: The image of the raw fingerprint is captured onto a silicon surface, and the resultant image is then translated into electrical signals;
3. Ultrasound sensors: This is where acoustic signals are sent toward the finger, and then a receiver subsequently digitizes the echoes from the acoustic signals.

## The matching algorithm

As mentioned, it is the matching algorithm which compares the enrollment template with the verification template, and in order to ascertain the degree of similarity or closeness between the two, a certain methodology must be followed, and this is described as follows:

1. Whatever data is collected from the raw image of the fingerprint, it must have some sort of commonality with the enrollment biometric

template which is already stored in the database. This intersection of data is known as the core, which is also the maximum curvature in a ridgeline.

2. Any extraneous objects which could possibly interfere with the unique feature extraction process must be removed, before the process of verification/identification can actually occur. For example, some of these extraneous objects can be the various differences found in the size, pressure, and the rotation angle of the fingerprint, and these can be normalized and removed by the matching algorithm.
3. In the final stage, the unique features collected from the raw data (which actually becomes the verification template) must be compared to that of the enrollment template later. At this point, this is where the matching algorithm does a bulk of its work. The actual matching algorithm can be based upon the premise of three types and kinds of correlations:
   a. Correlation-based matching: When two fingerprints are overlaid among each other, or superimposed, differences at the pixel level are calculated. Although it is strived for, perfect alignment of the superimposed fingerprint images is nearly impossible to achieve. Also, a disadvantage with this correlation method is that performing these types of calculations can be very processing intensive, which can be a grave strain on computing resources.
   b. Minutiae-based matching: In fingerprint recognition, this is the most widely used type of matching algorithm. With this method, it is the distances and the angles between the minutiae which are calculated and subsequently compared with another. There is global minutiae matching as well as local minutiae matching, and the latter method focuses upon the examination of a central minutiae, as well as the nearest two neighboring minutiae.
   c. Ridge feature matching: With this matching method, the minutiae of the fingerprint are combined with other finger-based features such as shape and size, the number and the position of various singularities, as well as global and local textures. This technique comes to of great value if the raw image of the fingerprint is poor in quality, and these extra features can help compensate for that deficit.

## Fingerprint recognition – the advantages and disadvantages

Although fingerprint recognition is one of the dominant biometric technologies available today, it is very important to take an objective view of it, from the standpoint of its advantages and disadvantages. Remember, biometric technology is just that – just another piece of a security tool. It has its fair share of flaws just like anything else. Therefore, it is critical to examine these variables as well.

By looking at the advantages and disadvantages of any biometric system, as a C-Level Executive, you will be in a much better position in order to make the best procurement decision possible. The advantages and disadvantages of fingerprint recognition, as well as the other biometric technologies reviewed in this chapter, will be examined from seven different perspectives, and they are as follows:

1. Universality: Every human being has fingerprints. And, although a majority of the world's population can technically be enrolled into a fingerprint recognition system, a small fraction cannot, and thus, a manual system must be put into place in order to confirm the identity of these particular individuals;
2. Uniqueness: The uniqueness of the fingerprint is essentially written by the DNA code which we inherit from our parents. As a result, even identical twins are deemed to have different fingerprint structures. Interestingly enough, although the uniqueness of the fingerprint is accepted worldwide, there have been no concrete scientific studies to prove this hypothesis. So, the uniqueness of the fingerprint is still in theory only, unless it is supported by hardcore scientific tests and the results prove it;
3. Permanence: Although the basic features of a fingerprint do not change as we get older over time, the fingerprint is still very much prone to degradation from conditions which exist in the external environment, such as cuts and contact with corrosive chemicals;
4. Collectability: This refers to the collection of the actual raw image of a fingerprint. But unlike other biometric technologies (especially the non-contactless ones), fingerprint recognition can be affected by a number of key variables which can degrade the quality of the raw image(s) collected. These variables include moist, oily, and dirty fingers; excessive direct pressure (this is pressing the finger too hard on the sensor) as well as rotational pressure (this means that the finger rolls side to side on the sensor); dirt and oil left on the sensor; as well as residual fingerprint left by previous users (also known as latent fingerprints). Another disadvantage is that fingerprint recognition requires very close contact on part of the end user, thus rendering this technology almost useless for high level surveillance and identification (one to many) applications;
5. Performance: This is composed of two variables: Accuracy, and the ability to meet the specific requirements of the customer. The accuracy of a fingerprint recognition system is based upon such factors as the number of fingers which are used for both verification and identification applications; the quality and robustness

of the matching algorithms utilized; and the size of the database which holds the fingerprint biometric templates. Under this performance umbrella, fingerprint recognition offers a number of key advantages:

a. Due to the sheer number of vendors (the number of biometric vendors is among the largest in the biometric industry), the marketplace translates to much lower pricing when compared to the products and solutions of other biometric vendors;
b. Since fingerprint recognition is probably the most widely used biometric technology, this translates into easy training programs for the end user(s);
c. The size of the biometric template (both enrollment and verification templates) is very small, on the range of 250 bytes to 1 kilobyte;
d. Due to the advancements in fingerprint recognition technology, very small sensors have been created, thus allowing for this technology to be used on small portable devices like netbooks and smartphones.

6. Acceptability: This evaluation criterion involves to what degree fingerprint recognition can be accepted by society as a whole. While a majority of the world's population are accepting of fingerprint recognition, there are also serious objections to using it, which are described as follows:
   a. Privacy rights issues: This is where the use of fingerprint biometric templates falls outside of their intended uses, which further results in the degradation of the sense of anonymity;
   b. Concerns over hygiene: All fingerprint biometric systems require direct, human contact with the sensor. For example, the sensor can become contaminated with germs from other users of the system, which can be easily transmitted;
   c. There is a strong association with fingerprints and criminal activity. This is so because fingerprints have been the hallmark of law enforcement for the past number of centuries.
7. Resistance to circumvention: This refers to the point as to how well a fingerprint biometric system is hack proof. Meaning, to what degree can a biometric system be resistant to security threats and risks posed to it from outside third parties. While fingerprint recognition is mostly hacker proof, it does possess a number of vulnerabilities. Probably the biggest flaw is that of latent fingerprints which can be left behind in a sensor. As a result, these latent prints can be used to produce what is known as "dummy prints", and these could be used to possibly fool a biometric system. But also, keep in mind that while this scenario is a possibility, all fingerprint biometric systems require that a live fingerprint be read – one that is supported with

a temperature, electrical conductivity, and blood flow. This greatly reduced the risk of spoofing.

### Market applications of fingerprint recognition

As mentioned numerous times in these chapters, fingerprint biometrics is the most widely used biometric technology; therefore, it serves a wide number of market applications, and some of the following are just examples of its enormous breadth and depth:

1. Forensics: The most common example is that of capturing latent fingerprints;
2. The administration of government benefits: The goal here is to 100% verify the identity of people who are entitled to receive government benefits and entitlements, and to make sure that they are properly received;
3. Use in financial transactions: Fingerprint recognition is the most widely used for accessing cash at ATM machines worldwide, and allowing for cashless transactions. This works by associating the identity of a verified individual to that of a checking or savings account (or any other source of funds) from which money can be withdrawn, after the fingerprint is presented to the biometric system. A perfect example of this is the use of fingerprint recognition systems at grocery stores. For example, some grocery stores allow customers to pay for their products with a mere scan of their finger. The amount is then subsequently deducted from the customer's checking account which is tied to their biometric template;
4. Physical and logical access: This is probably the biggest market application for fingerprint recognition systems. Physical access refers to gaining access into buildings, or other secured areas from within inside a physical infrastructure. Logical access refers to gaining access to computer networks, corporate intranets, and computers and laptops themselves. In this regard, fingerprint recognition is the most widely used technology for single sign-on solutions (also known as SSOs). Meaning, instead of having to enter in a password, all one has to do is present their fingerprint to the device, and in less than one second that individual can be logged into their computer.

## HAND GEOMETRY RECOGNITION

Hand geometry recognition is older even than that of fingerprint recognition. This particular technology dates all the way back to the 1960s, it was patented in 1985, and by the mid-1990s, the evolution of fingerprint recognition started to take over the market dominance which was held by hand geometry recognition. Today, there is only one vendor which actually develops and manufactures the hand geometry scanners.

It should be noted that the hand itself does not possess any unique features per se, but rather it is the shape of the hand (or the geometry of the hand) which is unique among most individuals. Given this situation, hand geometry recognition is rarely used for large-scale identification purposes. Rather, it is used for verification purposes only.

And in this regard, it can serve very large-scale applications for very harsh environments, such as physical access entry for warehouses, storage facilities, and factories. It is also used widely in time and attendance scenarios as well. In these verification scenarios, very often a smart card is also used, which is used to store further encrypted data about the end user.

In the enrollment process, some 96+ measurements of the hand are taken, and this includes the following variables:

1. The overall shape of the hand;
2. The various dimensions of the palm;
3. The length and width of all ten fingers;
4. The measurement between the distances of the joints of the hand;
5. The various shapes of the knuckles;
6. The geometrical circumference of the fingers;
7. Any distinct landmark features which can be found on the hand.

## Hand geometry recognition – the enrollment process

To start the enrollment process with a hand geometry scanner, a combination of prisms and light emitting diodes (LEDs) from within the scanner is used, to help capture the raw images of the hand. The technology can capture images of the back of the hand as well as the palm. This creates a two-dimensional image of the hand.

In order to capture a three-dimensional image of the hand, five guiding pegs are located just underneath the camera, to help the end user position their hand properly. Although this is advantageous, this method also possesses one serious disadvantage. For instance, the images of these pegs are also captured into the raw image. This results in greatly increased processing time because the images of the pegs have to be removed.

Also, because of this, the extraction algorithm cannot take into account variances due to hand rotation and differences in the placement of the hand. To create the enrollment template, the average of the measurements described previously are calculated, which is then converted into binary mathematical file. This is very small, only 9 bytes.

A problem in the construction of the enrollment and verification templates is that the geometric features of the hand share quite a bit of resemblance or correlation with another, which can greatly hinder the process of unique feature extraction. To help alleviate this problem, a method known as principal component analysis, or PCA, is used to produce a set of geometric features which are uncorrelated, and thus, unique features can be extracted.

The small biometric template size of hand geometry scanning gives it a rather distinct advantage. For example, hand geometry recognition works very effectively and is very interoperable as a multimodal security solution for both physical and logical access entry, as well as for time and attendance applications. As a standalone device, a hand geometry recognition device can store upward of 40,000+ unique biometric templates.

## Hand geometry recognition – the advantages and disadvantages

As reviewed in the strengths and weaknesses of fingerprint recognition, the same seven criteria can be used to review the advantages and disadvantages of hand geometry recognition as well:

1. Universality: When compared to other biometric technologies on a spectrum, hand geometry recognition does very well here. This is so because most end users have at least one hand which they can use in order to be verified. The technology has advanced to the point that it can take into account, up to a certain degree, any deformities of the hand. Also, hand geometry recognition is not affected at all by any type or kind of skin color – it is just the shape of the hand which is captured. Also, the technology is very easy to use and train individuals on. For example, at most, an end user needs to know how to place their hand properly onto the platen, and how to swipe a smart card;
2. Uniqueness: Although each individual does possess a different hand shape, the unique features which are extracted do not possess the very rich information and data like the iris and retina possess;
3. Permanence: Although the dimensions of the hand are relatively stable throughout the lifetime of an individual, the hand is very susceptible to physical changes. This includes such factors as weight loss and weight gain, injuries, as well as various diseases;
4. Collectability: This is another key advantage for hand geometry recognition. Except for the variables just described in point 3, the raw images of the hand which are captured are not affected at all to things which could affect the surface of the skin, and this includes primarily grime, dirt, and scars. Thus, this key strength makes hand geometry recognition very suitable for very harsh and extreme weather environments;
5. Performance: Overall, hand geometry scanners are very easy to use, and its very small template size (only 9 bytes) makes this technology

very interoperable with other types and kinds of security systems. Also, the accuracy rate of hand geometry recognition is very high, ranked with an equal error rate (ERR) of 0.1% (see Chapter 1 for a detailed explanation of the ERR, as well as other important performance metrics). But hand geometry scanners are large and bulky, and this can prove to be a very serious disadvantage;
6. Acceptability: For the most part, hand geometry recognition is well received by the end user population and is perceived to be very non-invasive. The only major in this aspect is the hygiene of the hand geometry scanner, such as any germs or contaminants left behind by other users;
7. Resistance to circumvention: Just because a hand geometry scanner is very easy to use, on the contrary, it is very difficult to spoof. Trying to fool this kind of system would mean creating an entire three-dimensional hand, using a false, physical mock-up of it.

As mentioned, hand geometry recognition is used primarily for physical access entry and time/attendance applications. But it has also made its way at the international airport setting, used primarily in the US Citizenship and Immigration Service Passenger Accelerated System, or also known as INSPASS.

This program has been designed so that frequent flyer passengers, especially those of business travelers, can be enrolled into what is known as a "trusted travel program". With this, a passenger can enter into the country of their destination with a mere scan of their hand, rather than wasting time at customs and immigration lines using the traditional paper passport. Currently, this program is in use in the United States, Canada, and Israel.

## Vein pattern recognition

Vein pattern recognition is also known as "Vascular Pattern Recognition", or simply "VPR" for short. Out of all of the biometric technologies reviewed in this chapter, this technology is the most up and coming in the marketplace and is giving very serious competition to fingerprint recognition and to a certain degree, even iris recognition.

VPR is considered to be what is known as an automated physical biometric. This simply means that no direct contact is required by the end user on a VPR system. So, this classifies VPR as a non-contactless biometric technology, thus giving it its great appeal to the public and customers at large. VPR examines the unique pattern of blood vessels in either the palm or the finger (in this case, multiple fingers can be used).

A question which often gets asked at this point is why are the veins chosen for this particular technology over the arteries? The answer is quite simple, for three primary reasons:

1. Veins are much larger than the arteries;
2. Veins are much closer to the surface of the skin, the arteries are located much below the skin surface;
3. Veins carry deoxygenated blood throughout the entire body.

It is the third feature of the physiology and the anatomy of the veins which form the science behind VPR. For example, in order to extract the unique features of the vein pattern, a near infrared (NIR) light is exposed to either the palm or the fingertips. The hemoglobin absorbs the NIR light. Hemoglobin is merely a pigment in the bloodstream which is heavily constituted of iron and carries oxygen in the bloodstream. It is a fact that the veins in our body which contain hemoglobin have a much lesser concentration of it than the arteries which exist in our body.

An illustration of vein pattern recognition can be seen in Figure 2.2.

As a result of this, the veins absorb much more of the NIR light, and thus, they appear much darker and robust when the raw images are captured. At this point, one may question where the veins get their unique pattern

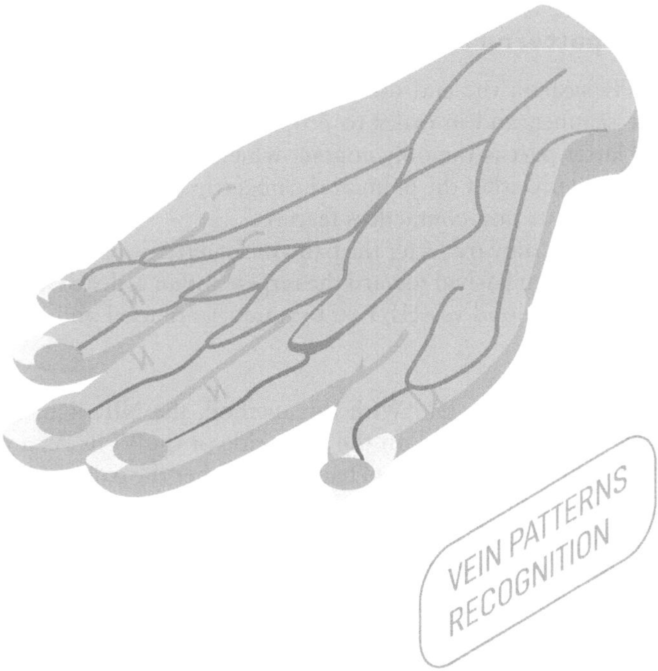


*Figure 2.2* This is an image of the veins in the hand, showing the unique features that can be extracted. (Biometric Authentication Vein Pattern Recognition on Stock Vector (Royalty Free) 2101294333 | Shutterstock.)

from. Just like the unique features of the iris and the fingerprint, the unique pattern of veins is created and formed from within the first eight weeks of the gestation cycle, and hence, this unique vascular pattern is then set from the moment we are born.

## The components of vein pattern recognition

The actual VPR technology is composed of four major components, which are as follows:

1. A sensor with LEDs which emit the NIR light as previously described;
2. A high-resolution charged-coupled device (CCD) camera;
3. A processing unit which extracts the unique pattern of blood vessels from either the palm or the fingertips, and which also produces the enrollment and verification templates;
4. The database which contains the enrollment and verification templates, as well as a detailed log history of the verification and identification transactions which have transpired during a set period of time.

## How vein pattern recognition works

With a VPR system, the end user can place either their palm or fingertip within the scanner, and in order to properly position them, guides can be used. The latter part is the only contact which is required, and there is no direct contact with either the palm or the fingertip on the actual sensor itself, thus giving VPR its non-contactless feature.

From this point onward, after the palm or fingertip is properly positioned, the NIR light is then flashed toward the target, and in order to illuminate the unique pattern of blood vessels, two techniques are used and are described as follows:

1. Diffused Illumination: With this method, the NIR light source and the sensor are located on the same side of the VPR. As a result, the NIR light which is reflected back from either the palm or the fingertip is then used to capture the raw vein pattern. The thickness of the skin is not a factor with this method;
2. Direct Illumination: With this technique, the NIR light is flashed directly at either the palm or the fingerprint, and through the target. This gives the impression of the ability to see the vein patterns right through the skin. This technique is directly affected by the thickness of the skin. The NIR light can be flashed from either the top, below, or on the sides of the VPR scanner. All three of these have their own set of advantages and disadvantages, which are as follows:

a. Top lighting: This gives the most robust raw image of the vein patterns, but these lighting units can be large as well as be exposed very quickly to dirt and grime;
b. Side lighting: This requires extra NIR lighting power as well as much more processing power to capture the raw image of the vein pattern;
c. Bottom lighting: This is the cheapest option, but it is much more sensitive to the actual NIR light, and for capturing the raw images of the vein pattern originating from a fingertip.

From within the VPR system itself, specialized software is used to normalize the raw images of the vein pattern which has just been captured and also removes any types obstructions such as skin hair on the palms or on the fingertips. The extraction algorithms then extract the unique features of the vein pattern, and the enrollment template remains encrypted in the database of the VPR system.

## Vein pattern recognition – the advantages and disadvantages

Just like fingerprint recognition and hand geometry recognition, VPR can also be rated and compared against the same seven criteria:

1. Uniqueness: Scientific studies have shown that even identical twins have unique vein patterns and structures;
2. Universality: VPR technology has an acceptance rating of very near 100% and is very versatile to use, when compared to the other biometric technologies. For instance, as described, either the palm or multiple fingers can be used, and if need be, even a combination of both the palm and the fingers can be used, thus providing for a very effective multimodal biometric solution;
3. Permanence: Questions are still being raised if the vein pattern does indeed change in the structure and physiology over the lifetime of an individual, although at the present time, it poses no serious issues for the purposes of verification and identification with VPR technology;
4. Collectability: VPR technology is very easy to use, manage, as well as deploy. Verification is very quick (less than 1 second), and since this is a non-contactless technology, the raw images of the vein are not at all impacted by dirt on the surface of the skin, cuts, bruises, scars, or even moisture and dryness on the fingertips or the palm. The only variable which can impact the quality of the vein pattern raw image is the thickness of the skin itself;
5. Performance: VPR possesses two very distinct advantages over the traditional biometric technologies:

   a. The matching algorithms are not at all complex and are also very "compact" in nature; thus, the processing and storage demands toward a VPR system are quite low;
   b. The sensor technology has made great advancements, and hence, it has become very small in nature, thus positioning VPR technology as a strong potential to be used for smartphone security;
   c. It should be noted, though, that a VPR system can be very prone to the negative effects of ambient light, or light which comes from the external environment;
6. Acceptability: Because of its non-contactless nature, VPR technology is viewed much more as a "hygienic" form of biometric technology (especially in countries like Japan), thus giving VPR its great appeal to customers. There are also virtually no privacy rights or civil liberties issues with VPR technology, because an individual's vascular pattern is basically invisible to the outside world;
7. Resistance to circumvention: It is very difficult to spoof a VPR system, because just like the other biometric technologies, a VPR system requires a live subject, because a constant blood flow in the veins is required in order for the raw images to be captured.

Given how well VPR ranks against these seven criteria, as well as its wide public acceptance, it is just a given fact that VPR serves a wide range of markets, as well as specialized markets. The applications that VPR serves are both government and commercial applications. Examples of such include the following:

1. Confirming the identity of government welfare program recipients;
2. Using VPR as an SSO for logical access entry, thus eliminating the use of passwords entirely;
3. VPR is also seeing heavy usage as an SSO at point-of-sale (POS) terminals;
4. Wide scale usage in physical access entry scenarios (primarily for office building type of entry, VPR is not meant for the harsh environments like warehouses and factories – this is where hand geometry recognition comes into play);
5. VPR is also seeing a huge uptick in the usage at the major international airports as a way of quickly and efficiently confirming the identity of passengers via the use of their e-Passport;
6. VPR is an excellent choice for time/attendance applications for all kinds and types of workforces;
7. VPR is even being used at car rental places, mini storage units, as well as condominiums and dormitories in universities and colleges.

## Palm print recognition

Another biometric technology which relates to that of hand geometry recognition is that of palm print recognition. There is a great deal of confusion between vein pattern recognition and palm print recognition, in that the two are very similar. However, the two are actually quite different from each other. While one can measure the unique pattern of blood vessels from the palm itself, palm print recognition is just that: The recognition of the shape, or the geometry of the palm.

As a technology, it is probably the least used in the world of biometrics. But, just like the use of fingerprints, the images of the palm have been around for quite a long time. For example, dating all the way back to the British Civil Service, palm prints were used to confirm a worker's identity when it was time to receive their pay. The very first palm recognition system which was fully automated was manufactured and introduced in Hungary back in 1994, for use by law enforcement officials.

In terms of forensic purposes, palm prints play a very crucial role as latent evidence – in fact, experts claim that up to 30% of major crime scenes can contain latent palm prints, which are subsequently used later in a court of law. One of the biggest obstacles with palm recognition is that the palm is much larger than that of any inked or sensing surfaces.

This means that the image of the palm has to be captured from two entirely separate images. These images must then be overlapped with another and then literally "stitched" together to form the entire image of the palm. From this stitched image, the features of the palm can be very easily seen. These include the creases, ridge patterns, and the minutiae.

## How palm print recognition works

The science and technology behind palm print recognition is virtually the same as that of fingerprint recognition. The raw images of the palm are extracted, the templates are created, and the verification/identification takes place via the comparison of the enrollment templates versus the verification templates.

Palm print technology can either use optical, thermal, or even ultrasound sensors. The question often arises how unique the palm print is. While there have been no concrete scientific studies to prove the validity of any type or kind of uniqueness to be found in the palm, palm prints are created and formed in the womb and remain fairly stable after the individual is born.

But keep in mind also that the palm is very much prone to threats from the external environment, such as dirt, scars, injuries, grime, and the degradation as well as wearing down of its features over the lifetime of an individual.

### Palm print recognition: the advantages and disadvantages

Because palm print recognition is not widely used, it can be rated against only four of the performance criteria, and they are as follows:

1. Collectability: Because the palm consists of a much larger surface, there is a higher probability of collecting more unique features than other biometric techniques (such as fingerprint recognition, where the scan surface is obviously much smaller);
2. Performance: Palm print recognition works very well when it is used as a part of a multimodal security solution such as hand geometry recognition;
3. Acceptability: Palm print recognition can be prone to the social issues surrounding biometrics, because palm prints can be associated with certain types of diseases which include Down's syndrome, fetal alcohol syndrome, Aarskog syndrome, and Cohen syndrome;
4. Resistance to circumvention: One advantage of palm print recognition in this aspect is that it is much more difficult to spoof. For example, there are many features of the palm which have to be replicated, and a much larger "dummy" palm has to be fashioned.

## FACIAL RECOGNITION

Facial recognition is one of those biometric technologies which most people can associate with. For example, we all have a face, and just like the fingerprint, the face has been used to verify and identify criminals and wanted suspects, as well as terrorists. Probably one of the best examples of facial identification is the photos at the post office, as well as those facial images which are on the wanted portion of the major law enforcement websites.

But, unlike the other biometric technologies, being used today, facial recognition is very prone to privacy rights issues and claims of civil liberties violations. The primary reason for this is that facial recognition can be used very covertly, without the knowledge or the consent of the individuals of whom the system is trying to track down.

Also, facial recognition does have its fair share of technological flaws as well. For example, if a facial recognition system were to capture the image of an individual who is grossly overweight and then capture another image of the same person who went through massive weight loss, the facial recognition system would not be able to make a positive match. In other words, the system can be very easily spoofed in these aspects.

But it is not just weight loss which can trick a facial recognition system, but such other things as the presence and the subsequent removal of facial hair, aging, as well as the presence and absence of other objects on the face, such as hats, sunglasses, as well as the switching from contact lenses to eye glasses and vice versa. But one of the key advantages of facial recognition is that it can be used for both verification and identification scenarios, and for heavy duty usage as well.

For instance, facial recognition is used quite frequently in the e-Passport infrastructures of many nations around the world and is also used for large-scale identification applications at the major international airports, especially in hunting down those suspects on the terrorist watch lists.

## Facial recognition: how it works

Facial recognition technology relies upon the physical features of the face (see Figure 2.3) which are determined by genetics. Also, this technology can be deployed either as a fully automated system or as a semi-automated system. With the latter, no human interaction is needed; all of the verification and identification decisions are made by the technology itself. With the latter, human intervention to a certain degree is required, and this is actually the preferred method for deploying a facial recognition system.

Given some of the serious obstacles it still does face, it is always better to err on the side of caution and have an actual human being have involvement as well in rendering a verification or identification decision.

Facial recognition systems of today focus upon on those parts of the face which are not as easily prone to the hurdles just described. These regions of the face include the following:

*Figure 2.3* This is an image of the face, showing the unique features that can be extracted. (Future Face Detection Technological 3D Scanning Stock Photo 1628451379 | Shutterstock.)

1. The ridges between the eyebrows;
2. The cheekbones;
3. The mouth edges;
4. The distances between the eyes;
5. The width of the nose;
6. The contour and the profile of the jawline;
7. The chin.

The methodology to capture the raw images of the face is much different when compared to the other biometric technologies. Although facial recognition is a non-contactless technology, the image capture processes are much more complex, and more cooperation is required on part of the end user. To start the process of raw image collection, the individual must first either stand before a camera or, unknowingly, have their face captured with covert surveillance methods, such as using a CCTV camera system (with the technology that is available today, facial recognition can literally be implanted in a CCTV).

Once the raw images are collected by the camera, the data is then either aligned or normalized to help refine the raw images at much more granular level. The refinement techniques involved include adjusting the face to be in the middle of the pictures which have been taken, and adjusting the size and the angle of the face so that the best unique features can be extracted and later converted over to the appropriate verification and enrollment templates.

All of this is done via mathematical algorithms. As mentioned previously, facial recognition is countered by a number of major obstacles, but even more so at the raw image acquisition phase. These include a lack of subtle differentiation between the faces and other obstructive variables in the external environment, various different facial expressions and poses in subsequent raw image captures, and capturing a landmark orienting feature such as the eyes.

To help compensate for these obstacles, much research and development has been done in the area of what is known as three-dimensional imaging. In this technique, a shape is formed and created, and using an existing two-dimensional image, various features are created, and the resultant is a model which can be applied to any three-dimensional surface and can also be used to help compensate for the abovementioned differences.

However, it should be noted that these types of three-dimensional facial recognition systems are not widely deployed in commercial applications yet, because this technique is still considered to be under the research and development phases. Right now, it is primarily two-dimensional facial recognition systems which are used on the commercial market. Three-dimensional facial recognition systems are only used as a complement to the two-dimensional ones, in which higher imaging requirements are dictated, and the capture environment is much more challenging.

An illustration of facial recognition can be seen in Figure 2.3.

## Defining the effectiveness of a facial recognition system

According to the International Committee for Information Technology Standards, also known as INCITS, the raw image of a face must meet certain stringent requirements in order to guarantee its effectiveness and reliability in a facial recognition system. These requirements are as follows:

1. The facial raw image must include an entire composite of the head, the individual must possess a full head of hair, and the raw image should capture the neck and shoulders as well;
2. The roll, pitch, and yaw of the facial raw images collected must possess a variance of at least ±5 degrees of rotation;
3. Only plain and diffused lighting should be used to capture the facial raw images;
4. In order for verification and/or identification to take place, no shadows whatsoever should be present in the raw images collected.

If a three-dimensional facial recognition system is used, the following properties must be observed:

1. Stereo imaging must utilize at least two cameras, which are mounted at a fixed distance;
2. If structured lighting is used, the facial recognition system flashes a defined, structured pattern at the face, which is used to capture and compute depth;
3. Laser scanners are the most robust form of sensing but are very costly to implement as well as very slow; it takes as long as 30 seconds or even more to capture and process the raw image of a face;
4. Hybrid sensors do exist and can use both the stereo imaging and structured lighting techniques.

The entire process of facial recognition starts with the location of the actual image of a face within a set frame. The presence of the actual face can be sensed or detected from various cues or triggers, such as skin color, any type or kind of head rotation, the presence of the facial or even head shape, as well as the detection and presence of both sets of eyes in the face.

Some of the challenges involved in locating the face in the frame include identifying the differentiation between the tonality of the skin color and the background, the various shapes of the face (depending of course on the angle in which the raw image is actually presented to a facial recognition system), or even multiple images of faces may be captured into a single frame, especially if the facial recognition system has been used in a covert fashion, in a very large crowd.

## The techniques of facial recognition

To help alleviate these obstacles and to provide a solution in which a single facial image can be detected in just one frame, various techniques have been developed and applied to facial recognition. These techniques fall under two categories:

1. Appearance based;
2. Model based.

With appearance-based facial recognition techniques, a face can be represented in several object views, and it is based on one image only, and no three-dimensional models are ever utilized. The specific methodologies here include PCA, and linear discriminant analysis (LDA). Model-based facial recognition techniques construct and create a three-dimensional model of the human face, and after that point onward, the facial variations can be captured and computed. The specific methodology here includes Elastic Bunch Graph Matching. All of these techniques will now be discussed in greater detail.

With PCA (this is linear based), this technique dates all the way back to 1988, when it was first used for facial recognition. This technique primarily uses what is known as "Eigenfaces". Simply put, Eigenfaces are just merely two-dimensional spectral facial images, which are composed of grayscale features.

There are literally hundreds of Eigenfaces which can be stored in the database of a facial recognition system. When facial images are collected by the system, this library of Eigenfaces is placed over the raw images and is superimposed over one another. At this point, the level of variances between the Eigenfaces and the raw images is then subsequently computed, averaged together, and then different weights are assigned.

The end result is a one-dimensional image of the face, which is then processed by the facial recognition system. In terms of mathematics, PCA is merely a linear transformation in which the facial raw images get converted over into a geometrical coordinate system. Imagine, if you will, a quadrant-based system. With the PCA technique, the dataset with the greatest variance lies upon the first coordinate of the quadrant system (this is also termed the first PCA), the next dataset with the second largest variance falls onto the second coordinate, and so on, until the one-dimensional face is created.

The biggest disadvantages of this technique are that it requires a full frontal image, and as a result, a full image of the face is required. Thus, any changes in any facial feature require a full recalculation of the entire Eigenface process. However, a refined approach has been developed, thus greatly reducing the calculating and processing time which is required.

With LDA (this is linear based), the goal is to project the face onto a vector space, with the primary objective being to speed up the verification and identification processes by cutting down drastically on the total number of features which need to be matched.

The mathematics behind LDA is to calculate the variations which occur between a single raw data point from a single raw data record. Based from these calculations, the linear relationships are then extrapolated and formulated. One of the advantages of the LDA technique is that it can actually take into account the lighting differences and the various types of facial expressions which can occur, but still, a full face image is required.

After the linear relationship is drawn from the variance calculations, the pixel values are captured and statistically plotted. The resultant image is a computed raw image, which is just simply a linear relationship of the various pixel values. This raw image is called a Fisherface, which can be seen in. Despite the advantages a major drawback of the LDA technique is that it does require a large database.

With the Elastic Bunch Graph Matching (this is model based, also known as EBGM) technique, this looks at the nonlinear mathematical relationships of the face, which includes such factors as lighting differences, and the differences in the facial poses and expressions. This technique uses a similar technique which is used in iris recognition, known as Gabor wavelet mathematics.

With the EBGM technique, a facial map is created, and an example of this can be seen. The facial image on the map is just a sequencing of graphs, with various nodes located at the landmark features of the face, which include the eyes, edges of the lips, and tips of the nose. These edge features become two-dimensional distance vectors, and during the identification and verification processes, various Gabor mathematical filters are used to measure and calculate the variances of each node on the facial image.

Then, Gabor mathematical wavelets are used to capture up to five spatial frequencies, and up to eight different facial orientations. Although the EBGM technique does not at all require a full facial image, the main drawback with this technique is that the landmarks of the facial map must be marked extremely accurately, with great precision.

## Facial recognition: the advantages and the disadvantages

Facial recognition systems can also be evaluated against the same set of criterion. In this regard, there is a primary difference between this and the other biometric technologies. And that is, while the face may not offer the most unique information and data like the iris and the retina, facial recognition can be very scalable, and like fingerprint recognition and hand geometry recognition, facial recognition can fit into a wide variety of application environments.

The evaluation of facial recognition can be broken down as follows:

1. Universality: Unlike all of the other biometric technologies, every individual possesses a face (no matter what the condition of the face is actually in), so at least theoretically, it is possible for all end users to be enrolled into a facial recognition system;
2. Uniqueness: As mentioned, facial recognition is not distinctly unique at all, even members of the same family can genetically share the same types and kinds of facial features, as well as identical twins (when it comes to the DNA code, it is the facial features which we inherit the most resembling characteristics from);
3. Permanence: Given the strong effect of weight gain and weight loss (including the voluntary changes in appearance), as well as the aging process we all experience, permanence of the face is a huge problem. In other words, the face is not at all stable over time and can possess a large amount of variance. As a result, end users may have to constantly be re-enrolled in a facial recognition system time after time, thus wasting critical resources and processing power;
4. Collectability: The collection of unique features can be very difficult, because of the vast differences in the environment which can occur during the image acquisition phase. This includes the differences in lighting, lighting angles, and the distances at which the raw images are captured, and also including the extraneous variables such as sunglasses, contact lenses, eye glasses, and other types and kinds of facial clothing;
5. Performance: In this regard, facial recognition has both positive and negative aspects, which are as follows:
   - Accuracy: Facial recognition according to recent research has a false acceptance rate (FAR) of 0.001, and a false rejection rate (FRR) of 0.001;
   - Backward compatibility: Any type or kind of two-dimensional photograph can be added quite easily in the database of the facial recognition system, and subsequently utilized for identification and verification;
   - Lack of standardization: Many facial recognition systems are in existence, but there is a severe lack of standards among the interoperability of these systems;
   - Template size: The facial recognition biometric template can be very large, up to 3,000 bytes, and as a result, this can greatly increase the storage requirements as well as choke off the processing system of the facial recognition system;
   - Degradation: The constant compression and decompression and recycling of the images can cause serious degradation to the facial images which are stored in the database over a period of time;

6. Acceptability: In a broad sense, facial recognition can be widely accepted. However, when it is used for surveillance purposes, it is not at all accepted, because people believe that it is a sheer violation of privacy rights and civil liberties. Also, some cultures prohibit the use of facial recognition systems, such as the Islamic culture, where women are required to wear head scarves and hide their faces.
7. Resistance to circumvention: Facial recognition systems can be very easily spoofed and tricked by two-dimensional facial images.

## Applications of facial recognition

Given the covert nature and the ability to deploy to deploy facial recognition easily into other non-biometric systems and technologies, it is no wonder that it has a wide range of market applications it can serve. Probably the biggest application for facial recognition has been that for the e-Passport, and to a certain degree, the National ID Card system, for those nations who have adopted it.

For example, the International Civil Aviation Organization (also known as the ICAO) has made facial recognition templates the de facto standard for machine readable travel documents. Also, along with iris recognition, facial recognition is being used quite heavily at the major international airports, primarily for the covert surveillance purposes, in an effort to scan for individuals on the terrorist watch lists.

Also, contrary to public support of it, facial recognition can be used very covertly at venues where large crowds gather, such as sporting events and concerts. Facial recognition systems can also be used in conjunction with CCTV cameras and strategically placed in large metropolitan areas. For example, the City of London is a perfect example of this. At just about every street corner, there is a facial recognition/CCTV camera system deployed.

Because of this vast network of security over there, the London police were able to quickly apprehend and bring to justice the terrorist suspects who were involved in the train bombings which took place. Another popular application for facial recognition is for border protection and control, especially widely used in the European countries.

Facial recognition is also heavily used in conducting real-time market research studies. For instance, it can be used to gauge a potential customer's reaction to certain advertising stimuli by merely recording that individual's facial movements. Casinos are also a highly favored venue for facial recognition, as a means to identify and verify the welcomed guests versus the unwelcomed guests.

Facial recognition has also been used in both physical access entry and time and attendance scenarios, but nowhere near to the degree that hand geometry recognition and fingerprint recognition are currently being used in these types of market applications.

## The eye: the iris and the retina

Out of all of the biometric technologies available today, it is the eye, and its subcomponents, especially the retina and the iris, which possesses the most unique information and data in which to identify and verify an individual. As of now, it is the retina which is deemed to be the most unique biometric of all (of course, after the use of DNA which will be much further explored later in this chapter). The iris is deemed to be the next most unique biometric, right after the retina.

### THE IRIS

The popularity of iris recognition dates all the way back to the mid-1990s, when the first mathematical algorithms were developed and created by Dr. John Daugman of the University of Cambridge. At the time, only one iris recognition vendor existed, known as Iridian Technologies, Inc. For well over a decade, this company owned the intellectual property as well as the mathematical algorithms and the intellectual property involved with iris recognition at the time. As a result, a very strong monopoly was held, and customers could only utilize just one type of iris recognition product or solution.

The system, although considered to be very advanced at the time, was quite expensive and bulky to deploy and maintain. But eventually, just a couple of years ago, Iridian Technologies was bought out by a much larger biometrics vendor, and at the same time, the patents surrounding the original algorithms and technology also expired, thus breaking the monopoly which was held for such a long time.

As a result, there has been a huge explosion of iris recognition just in the past couple of years, offering many new types and kinds of products and solutions for the customer. The advantage to this of course is that the customer now has a much greater depth and breadth of picking and even customizing their own iris recognition solution which will fit their exact needs and requirements, unlike before, when people were basically forced to buy just one product or solution.

In fact, iris recognition has developed so quickly that now images of the iris can be captured at much greater distances, as well as when people are in movement. Previously, an end user had to stand directly in front of the iris recognition camera, at a very close proximity. And now, even the unique pattern of the blood vessels can also be scanned, and this will be examined after the section on retinal recognition.

## The physiological structure of the iris

The iris lies between the pupil and the white of the eye, which is known as the sclera. The color of the iris varies from individual to individual, but there is a commonality to the colors, and these include green, blue, brown, and in remote cases, even a hazel color can occur. And even in the most extreme cases, a combination of these colors can be seen in the iris. The color of the iris is primarily determined by the DNA code inherited from our parents.

The unique patterns of the iris start to form when the human embryo is conceived; usually this happens at about the third month of fetal gestation. The phenotype of the iris is shaped and formed in a process known as chaotic morphogenesis, and the unique structures of the iris are completely formed during the first two years of child development.

The primary purpose of the iris is to control the diameter and the size of the pupil. The pupil is the part of the eye which allows light to enter into the eye, which in turn reaches the retina, which is located in the back of the eye. Of course, the amount of light which can enter the pupil is a direct function of how much it can expand and contract, which is governed by the muscles of the iris. The iris is primarily composed of two layers: (1) A fibrovascular tissue known as the stroma, and (2) the stroma is in turn connected to a grouping of muscles known as the sphincter muscles.

It is this muscle which is responsible for the contraction of the pupil, and another group of muscles known as the dilator muscles governs the expansion of the pupil. When you look at your iris in the mirror, you will notice a

*Figure 2.4* This is an image of the iris, showing the unique features that can be extracted. (Iris Recognition Concept Smart Contact Lens Stock Photo 719253955 | Shutterstock.)

radiating pattern. This pattern is known as the trabecular meshwork. When NIR light is flashed onto the iris, many unique features can be observed. These features include ridges, folds, freckles, furrows, arches, crypts, coronas, as well as other patterns which appear in various, discernible fashions.

Finally, the collarette of the iris is the thickest region of it, which gives the iris its two distinct regions, known as the pupillary zone (this forms the boundary of the pupil), and the ciliary zone (which fills up the rest of the iris). Other unique features can also be seen in the collarette region. The iris is deemed to be one of the most unique structures of human physiology, and in fact, each individual has a different iris structure in both eyes. In fact, even scientific studies have shown that even identical twins have different iris structures. An image of the actual iris can be seen in Figure 2.4.

## Iris recognition: how it works

The idea of using the iris to confirm the identity of an individual dates all the way back to 1936, when an ophthalmologist, Frank Burch, first proposed the idea. This idea was then patented in 1987, and by the mid-1990s, Dr. John Daugman of the University of Cambridge developed the first mathematical algorithms for it. Traditional iris recognition technology requires that the end user stand no more than 10 inches away from the camera.

With the NIR light shined into the iris, various grayscale images are then captured and then compiled into one primary composite photograph. Special software then removes any obstructions from the iris, which can include portions of the pupil, eyelashes, eyelids, and any resulting glare from the iris camera.

From this composite image, the unique features of the iris (as described before) are then "zoned off" into hundreds of phasors (also known as vectors), whose measurements and amplitude levels are then extracted (using Gabor wavelet mathematics) and then subsequently converted into a binary mathematical file, which is not greater than 500 bytes. Because of this very small template size, verification of an individual can occur in just less than 1 second.

From within the traditional iris recognition methods, this mathematical file then becomes the actual iris biometric template, which is also known as the "IrisCode". However, in order to positively verify or identify an individual from the database, these iris-based enrollment and verification templates (the IrisCode) must be first compared with one another. In order to accomplish this task, the IrisCodes are compared against one another byte by byte, looking for any dissimilarities among the string of binary digits.

In other words, to what percentage do the zeroes and the ones in the iris-based enrollment and verification templates match up and do not match up against one another? This answer is found by using a technique known as

"Hamming Distances", which is even used in iris recognition algorithms of today.

After these distances are measured, tests of statistical independence are then carried out, using high-level Boolean mathematics (such as exclusive OR operators [XOR] and masked operators). Finally, if the test of statistical independence is passed, the individual is then positively verified or identified, but if the tests of statistical independence are failed, then the person is NOT positively verified or identified.

## The market applications of iris recognition

As a result of these technological breakthroughs in iris recognition, it now cuts across all realms of market applications. In the past decade, because of the monopolistic grip which held it, iris recognition served only a very limited number of market applications, and because of that, customer acceptance of it was very low, as well as the believability in the viability of this technology. Also, at that time, iris recognition technology was very expensive (about $3,000–$5,000), thus providing for a much bigger obstacle in its adoption rate. Today, the price has come down substantially, thus helping boost its acceptance by customers very quickly. Because of all of this, iris recognition has now become a dominant player in those market applications which were once traditionally held by hand geometry recognition and fingerprint recognition. These market applications include the following:

1. Logical access entry: This deals with gaining access to servers and corporate networks and intranets;
2. Physical access entry: This addresses issues of gaining access to secure areas in a facility or place of business;
3. Time and attendance: This deals with accurate time reporting and actual hours worked by employees;
4. Critical infrastructures: This includes nuclear power plants, oil refineries, large-scale military installations, as well as government facilities;
5. Airports: This includes everything from confirming a passenger's identity as they make their way through the security checkpoints to reading their e-Passports before they disembark at the country of their destination;
6. Seaports: This involves all of the maritime activities which include securing marine terminals, and "high consequence" facilities such as oil and gas storage, chemical, intermodal, and port operations;
7. Military checkpoints: This includes securing, with iris recognition, such areas as military bases, air force bases, and naval bases.

## Iris recognition: the advantages and the disadvantages

Iris recognition, when evaluated against the seven criteria, actually fares very well in terms of both acceptance and applicability for the markets it serves:

1. Universality: In a general sense, everybody has at least one eye which can be scanned and the unique features extracted from, so literally, the technology can be used anywhere around the world. Even in the unfortunate chance that an individual is blind in one or both eyes, technically, the iris can still be scanned, even though it can be much more difficult for the iris recognition system to scan the actual image of the iris;
2. Uniqueness: As mentioned, the iris, along with the retina, is one of the richest biometrics in terms of unique information and data. Even among identical twins, the irises are also unique. For example, Dr. John Daugman, who originally developed the iris recognition algorithms, calculated that the statistical probability of the iris being identical among twins is 1 in $10^{78}$;
3. Permanence: One of the biggest advantages of the iris is that it is very stable, and in fact, it hardly changes over the lifetime of an individual. Thus, in this regard, iris recognition does not possess the same flaws as facial recognition does. Also, the iris is considered to be an "internal" organ; thus, it is not prone to the harsh conditions of the external environment, unlike the face, hand, or fingers. However, it should be noted that any direct, physical injury to the iris, or even a laser-based vision corrective surgery, will render the iris useless;
4. Collectability: When compared to all of the other biometric technologies, the images of the iris are extremely easy to collect, both in terms of camera capture and the software analysis of the iris. In fact, the images of the iris can still be captured even when an individual is still wearing their eye glasses or contact lenses. The only disadvantage of iris recognition in this particular regard is that the image of the iris can be very much prone to blurring by the capture camera, and any excessive head tilts on the part of the individual can cause the iris image to be greatly distorted;
5. Performance: Given the very small size of the iris recognition template, this technology appears to be extremely fast as well as extremely accurate. For example, it possesses an FAR level of 0.0001, and an FRR at a remarkable level of 0.0. As a result of these metrics, iris recognition works very well for identification-based applications, as this is best exemplified by the iris recognition deployments in both Iraq and Afghanistan, where the technology is being used every day to identify hundreds of refugees in those particular countries;

6. Acceptability: Because iris recognition is non-contactless technology, overall, the acceptance has been pretty high. However, initially, there can be some unease and fear of having one's iris scanned. It should be noted that the NIR light used in iris recognition is not visible to humans, and there has been no physical harm cited as of yet to the eye, as a result of using iris recognition;
7. Resistance to circumvention: Iris recognition technology is very difficult to spoof, because the system can discriminate a live iris from a fake iris by carefully examining the dilation and constriction of the pupil, and the use of encryption to further protect the iris recognition templates greatly prevents against reverse engineering.

## CASE STUDY 1: "THE AFGHAN GIRL – SHARBAT GULA"

Probably the best example of using these traditional iris recognition techniques was in trying to confirm the identity of Sharbat Gula, also known as the "Afghan Girl". The famous picture of the striking, green eyes of Sharbat Gula goes back for almost 30 years, to 1985. She is originally from Afghanistan. During the time of the Soviet Union's invasion of Afghanistan, she and her family fled into the Nasir Bagh refugee camp, located in Peshawar, Pakistan.

During this time, a world-famous photographer from *National Geographic* magazine, Steve McCurry, came to this region to capture the plight of the Afghanistan refugees on pictures. One of the refugees he met and took pictures of was Sharbat Gula, when she was only 13 years old. Steve McCurry did not know her name, or anything else about her family. Of the many pictures he took of her, one stood out in particular – her piercing, green eyes.

During the course of the next 17 years, this picture became world famous, appearing in books, magazines, newspapers, posters, and other forms of media. Sharbat Gula knew nothing of her fame until she met Steve McCurry for the second time in January 2002. In January 2002, Steve McCurry returned to the same region in a final attempt to locate Sharbat Gula.

Steve McCurry and his team searched through numerous villages and came across various leads which proved to be false. Finally, the break came when an individual came forward and claimed that Sharbat Gula was a next-door neighbor many years ago. After several days of making this claim, this same individual brought back the brother of Sharbat Gula, who had the same color eyes. From that moment onward, McCurry and his team felt that they had located the family of Sharbat Gula.

Because of her culture, Sharbat Gula was not allowed to meet other men. However, a female producer of *National Geographic* was allowed to initially meet with and take photographs of her. Finally, after a series of "negotiations" with her family, Steve McCurry was able to see Sharbat Gula.

After asking some questions and comparing the world-famous photograph to photographs just taken of her, Steve McCurry felt he had finally found the Afghan Girl – 17 years later.

However, various tests had to be conducted – in order to make sure that Sharbat Gula was truly the Afghan Girl. Numerous tests were conducted to confirm the identity of Sharbat Gula. Two sophisticated tests were utilized: Facial recognition techniques developed by forensic examiners at the FBI, and iris recognition techniques developed by Dr. John Daugman and Iridian Technologies, Inc.

The pictures taken in 1985 were compared to the pictures taken in 2002, in both tests. The facial recognition techniques confirmed her identity; however, the ultimate test came down to iris recognition because of its reliability, as stated before in this chapter. However, the scientists at Iridian Technologies, Inc. had to overcome a number of obstacles.

First, the pictures taken by *National Geographic* of Sharbat Gula were not taken by iris recognition cameras, rather they were taken with other types of cameras. As a result, the scientists had to eliminate the effects of the light reflections produced by these cameras and also make various modifications to the image quality of these photographs.

The pictures of Sharbat Gula were then eventually scanned into a digital format. Second, iris recognition works by examining scans from live subjects, not static photographs – another major obstacle. After making a series of adjustments to the iris recognition software, the scientists concluded that Sharbat Gula was positively the Afghan Girl:

> The match of Sharbat Gula's eyes to the eyes on the 1985 cover photo was irrefutable, as we achieved a 1 in 100 million probability of a false match. There is no doubt in my mind that National Geographic has found the girl in the cover photo.
>
> *Quote from Ulf Cahn von Seelen, Director of Algorithms, Iridian Technologies, Inc., taken from Biometric Digest, April 2002*

## THE RETINA

The next component of the eye to be examined is the retina. Very often, there is a lot of confusion between the retina and the iris. As discussed in length, the iris is located in the front of the eye, but the retina is located at the back of the eye. Unlike the iris, the retina is actually a grouping of blood vessels which lead into the optic disc, and from there, visual information is transmitted to the brain for processing via the optic nerve, which lies in between the optic disc and the brain.

An image of the retina can be seen in Figure 2.5.

The scanning of the retina is also known as retinal recognition and, at the present time, is deemed to be the most unique biometric of all because of

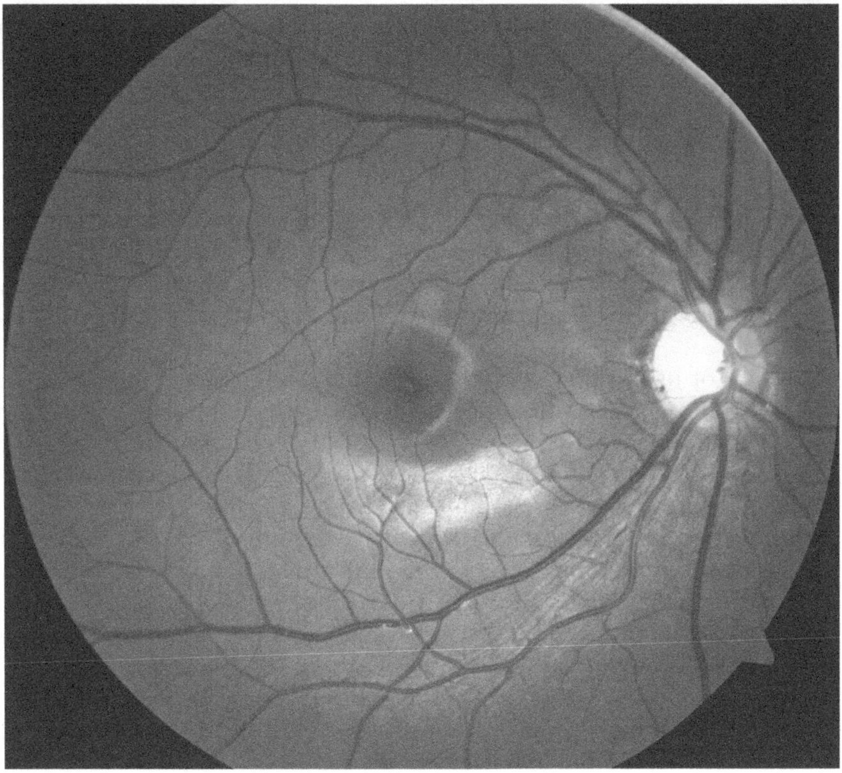

*Figure 2.5* This is an image of the retina, showing the unique features that can be extracted. (Retina Human Eye Stock Photo 2492524165 | Shutterstock.)

the richness of the data it possesses, from which the unique features can be extracted from.

Retinal recognition technology was first conceived of in the mid-1970s, but the first retinal recognition biometric device did not come out onto the marketplace until the early 1980s. Although retinal recognition does possess some very distinct advantages over the other biometric technologies, this technology has extremely limited market applications.

The primary reasons for this are the high cost of retinal recognition devices (which can be as high as $5,000 per device), its bulkiness, and the fact that it is very user invasiveness. Although retinal recognition technology is a non-contactless technology, the end user has to sit very close to the scanner and literally place his or her eye onto the scanning device, thus making it very cumbersome and very uncomfortable for the end user. An image of the retina can be seen in Figure 2.4, and overall side view of the eye, which includes both the retina and the iris, can be seen in Figure 2.5.

## The physiology of the retina

It is said that the retina "is to the eye as film is to a camera".[3] The retina consists of multiple layers of sensory tissue and millions of photoreceptors whose function is to transform light rays into electrical impulses. These impulses subsequently travel to the brain via the optic nerve, where they are converted to images. Two distinct types of photoreceptors exist within the retina: the rods and the cones.

While the cones (of which each eye contains approximately 6 million) help us to see different colors, the rods (which number 125 million per eye) facilitate night and peripheral vision. It is the blood vessel pattern in the retina that forms the foundation for retinal recognition as a science and technology.

Because of its position within the eye, the retina is not exposed to the external environment. As a biometric, it is therefore very stable. It is from here that information is transmitted to and received from the brain. The circle in the diagram indicates the area that is typically captured by a retinal scanning device. It contains a unique pattern of blood vessels.

There are two famous studies that have confirmed the uniqueness of the blood vessel pattern found in the retina. The first was published by Dr Carleton Simon and Dr Isadore Goldstein in 1935 and describes how every retina contains a unique blood vessel pattern. In a later paper, they even suggest using photographs of these patterns as a means of identification. The second study was conducted in the 1950s by Dr Paul Tower. He discovered that – even among identical twins – the blood vessel patterns of the retina are unique and different.

The first company to become involved in the research, development, and manufacture of retinal scanning devices was EyeDentify Inc. The company was established in 1976 and its first retina capturing devices were known as "fundus cameras". While intended for use by ophthalmologists, modified versions of the camera were used to obtain retina images. The device had several shortcomings, however. First, the equipment was considered very expensive and difficult to operate.

Second, the light used to illuminate the retina was considered too bright and too discomforting for the user. Further research and development yielded the first true prototype scanning device, which was unveiled in 1981. The device used infrared light to illuminate the blood vessel pattern of the retina. The advantage of infrared light is that the blood vessel pattern in the retina can "absorb" such light much faster than other parts of the eye tissue.

The reflected light is subsequently captured by the scanning device for processing. In addition to a scanner, several algorithms were developed for the extraction of unique features. Further research and development gave birth to the first true retinal scanning device to reach the market: The EyeDentification System 7.5. The device utilized a complex system of scanning optics, mirrors, and targeting systems to capture the blood vessel pattern of the retina.

Ongoing development resulted in devices with much simpler designs. Later scanners consisted of integrated retinal scanning optics, which sharply reduced manufacturing costs (compared to the EyeDentification System 7.5.).

The last retinal scanner to be manufactured by EyeDentify was the ICAM 2001, a device capable of storing up to 3,000 templates and 3,300 transactions. The product was eventually withdrawn from the market on account of its price as well as user concerns.

## The process of retinal recognition

The overall retinal scanning process may be broken down into three sub-processes:

1. Image/signal acquisition and processing: This sub-process involves capturing an image of the retina and converting it to a digital format;
2. Matching: A computer system is used to verify and identify the user (as is the case with the other biometric technologies reviewed in previous articles);
3. Representation: The unique features of the retina are presented as a template.

The process for enrolling and verifying/identifying a retinal scan is the same as the process applied to other biometric technologies (acquisition and processing of images, unique feature extraction, template creation).

The image acquisition and processing phase is the most complicated. The speed and ease with which this sub-process may be completed largely depends on user cooperation. To obtain a scan, the user must position his/her eye very close to the lens.

To safeguard the quality of the captured image, the user must also remain perfectly still at this point. Moreover, glasses must be removed to avoid signal interference (after all, lenses are designed to reflect). On looking into the scanner, the user sees a green light against a white background. Once the scanner is activated, the green light moves in a complete circle (360 degrees). The blood vessel pattern of the retina is captured during this process.

Generally speaking, three to five images are captured at this stage. Depending on the level of user cooperation, the capturing phase can take as long as 1 minute. This is a very long time compared to other biometric techniques. The next stage involves data extraction. One very considerable advantage of retinal recognition becomes evident at this stage.

As genetic factors do not dictate the pattern of the blood vessels, the retina contains a diversity of unique features. This allows up to 400 unique data points to be obtained from the retina. For other biometrics, such as fingerprints, only 30–40 data points (the minutiae) are available.

During the third and final stage of the process, the unique retina pattern is converted to an enrollment template. At only 96 bytes, the retina template is considered one of the smallest biometric templates.

As is the case with other biometric technologies, the performance of the retinal scanning device may be affected by a number of variables, which could prevent an accurate scan from being captured. Poor quality scans may be attributable to the following:

1. Lack of cooperation on the part of the user: As indicated, the user must remain very still throughout the entire process, especially when the image is being acquired. Any movement can seriously affect lens alignment;
2. The distance between the eye and the lens is incorrect and/or fluctuates: For a high-quality scan to be captured, the user must place his or her eye in very close proximity to the lens. In this sense, iris scanning technology is much more user-friendly; a quality scan can be captured at a distance of up to 3 feet from the lens;
3. A dirty lens on the retinal scanning device: This will obviously interfere with the scanning process;
4. Other types of light interference from an external source;
5. The size of the user's pupil: A small pupil reduces the amount of light that travels to (and from) the retina. This problem is exacerbated if the pupil constricts as a result of bright lighting conditions, which can result in a higher false reject rate.

## The advantages and disadvantages of retinal recognition

All biometric technologies are rated against a set of performance standards. As far as retinal recognition is concerned, there are two performance standards: The false reject rate, and the Ability to Verify Rate.

As mentioned back in Chapter 1, the false reject rate (FAR) describes the probability of a legitimate user being denied authorization by the retinal scanning system. Retinal recognition is most affected by the false reject rate.

This is because the factors described above have a tangible impact on the quality of the retinal scan, causing a legitimate user to be rejected. Also, the Ability to Verify Rate describes the probability of an entire user group being verified on a given day. For retinal recognition, the relevant percentage has been as low as 85%. This is primarily attributable to user-related concerns and the need to place one's eye in very close proximity to the scanner lens.

The strengths and weaknesses of retinal recognition: Just like all other biometric technologies, retinal recognition has its own unique strengths and weaknesses. The strengths may be summed up as follows:

1. The blood vessel pattern of the retina rarely changes during a person's life (unless he or she is afflicted by an eye disease such as glaucoma and cataracts);
2. The size of the actual template is only 96 bytes, which is very small by any standards. In turn, verification and identification processing times are much shorter than they are for larger files;
3. The rich, unique structure of the blood vessel pattern of the retina allows up to 400 data points to be created;
4. As the retina is located inside the eye, it is not exposed to (threats posed by) the external environment. For other biometrics, such as fingerprints and hand geometry, the opposite holds true.

The most relevant weaknesses of retinal recognition are as follows:

1. The public perceives retinal scanning to be a health threat; some people believe that a retinal scan damages the eye;
2. User unease about the need to position the eye in such close proximity of the scanner lens;
3. User motivation: Of all biometric technologies, successful retinal scanning demands the highest level of user motivation and patience;
4. Retinal scanning technology cannot accommodate people wearing glasses (which must be removed prior to scanning);
5. At this stage, retinal scanning devices are very expensive to procure and implement.

As retinal recognition systems are user invasive as well as expensive to install and maintain, retinal recognition has not been as widely deployed as other biometric technologies (particularly fingerprint recognition, hand). To date, retinal recognition has primarily been used in combination with access control systems at high-security facilities.

This includes military installations, nuclear facilities, and laboratories. One of the best-documented applications involves the State of Illinois, which used retinal recognition to reduce welfare fraud by identifying welfare recipients (thus preventing multiple benefit payments). This project also made use of fingerprint recognition.

Retinal recognition was first introduced in Granite City and East Alton (southern Illinois) toward mid-1996. Once the fingerprint recognition program had been initiated, the authorities drew up a comparison between fingerprint and retinal recognition, concluding that retinal scanning is not client or staff friendly and requires considerable time to secure biometric records. Based on these factors, retinal scanning technology is not yet ready for state-wide adaptation to the Illinois welfare department. As a result, the use of retinal recognition systems was stopped.

Retinal recognition can also be compared to the seven criteria the other technologies have been compared to. But while retinal recognition does possess some very significant advantages, overall, it does not fare as well (meaning the negatives far outweigh the positives) which seriously limits and curtails its adoption rate in the marketplace.

1. Universality: Virtually everybody, unless they have some extreme form of blindness, possesses a retina; thus, at least on a theoretical level, people can have their retina scanned;
2. Uniqueness: As described, the retina is very unique, even among identical twins. This is probably its greatest strength;
3. Permanence: Unlike the other physiological components of our bodies, the biological fundamentals of the retina hardly change in the lifetime of the individual. But it should be noted that if anything affects the iris, then the retina can be subject to degradation as well. Also, it should be noted that the retina is prone to degradation as well via diabetes, glaucoma, high blood pressure, and even heart disease;
4. Collectability: The scan area of the retina is very small, and the end user must place their eye in a very user invasive eye receptacle, and also, he or she must remove their contact lenses or eye glasses in order for a quality scan to be captured. Because of this very constrained environment, the end user must remain cooperative, and the systems administrator must make sure that the retina is scanned in an appropriate time frame; if not, the FTE (failure to enroll rate) will be high;
5. Performance: As a flip side to the user invasiveness which is involved, retinal recognition has extremely high levels of accuracy; in fact, it is claimed that the error rate is as low as one in 1 million;
6. Acceptability: Retinal recognition works best with those end users which are absolutely required to use it in order to perform their required job functions;
7. Resistance to circumvention: It is almost impossible to spoof a retinal recognition system, and a live retina is required for either verification or identification to take place.

Because of the high expense and user invasiveness involved, retinal recognition is only used for ultra-high-security applications, such as for the government and the military. As a result, there is a high level of preference given to iris recognition over retinal recognition.

## VOICE RECOGNITION

Voice recognition is a biometric technology whose research and development dates all the way back to World War II. For example, at that time, spectrographs showed that there are variations in the intensity of various sounds in a person's voice, and at different frequency levels.

This propelled the idea of perhaps using voice recognition to confirm the identity of a particular individual. The research and development into voice recognition continued well into the 1960s, and the voice spectrographs which were used at the time started to utilize statistical modeling as a means of biometric template creation, rather than using the traditional approaches.

This continued trend would allow for the evolution of automated voice recognition tools to come into play. In fact, the first known voice recognition system was called the "Forensic Automatic Speaker Recognition", or FASR for short.

In today's biometric world, voice recognition can be considered to be both a behavioral-based and a physical-based biometric. This is so because the acoustic property of a particular person's voice is a direct function of the shape of the individual's mouth, as well as the length and the quality of the vocal cords (the physical component). But also, at the same time, the behavioral data of an individual's voice is present in the template as well, and this includes such variables as the pitch, volume, and the rhythm of the voice.

An image of voice recognition can be seen in Figure 2.6.

*Figure 2.6* This is an image of the voice, showing the unique features that can be extracted. (Girl Using Smartphone Voice Recognition Stock Photo 552056272 | Shutterstock.)

## Voice recognition – how it works

The first step in voice recognition is for an individual to produce an actual voice sample. Voice production is a facet of life which we take for granted every day, and the actual process is complicated. The production of sound originates at the vocal cords. In between the vocal cords is a gap. When we attempt to communicate, the muscles which control the vocal cords contract.

As a result, the gap narrows, and as we exhale, this breath passes through the gap, which creates sound. The unique patterns of an individual's voice are then produced by the vocal tract. The vocal tract consists of the laryngeal pharynx, oral pharynx, oral cavity, nasal pharynx, and the nasal cavity. It is these unique patterns created by the vocal tract which is used by voice recognition systems.

Even though people may sound alike to the human ear, everybody, to some degree, has a different or unique annunciation in their speech. To ensure a good quality voice sample, the individual usually recites some sort of text, which can either be a verbal phrase a series of numbers or even repeating a passage of text, which is usually prompted by the voice recognition system. The individual usually has to repeat this a number of times.

The most common devices used to capture an individual's voice samples are computer microphones, cell (mobile) phones, and the land line-based telephones. As a result, a key advantage of voice recognition is that it can leverage existing telephony technology, with minimal disruption to an entity's business processes. In terms of noise disruption, computer microphones and cell phones create the most, and land line-based telephones create the least.

## Factors affecting voice recognition

But it is very important to note at this point that the same medium which was used to enroll the end user into the voice recognition system uses the same type of medium for later uses. For example, if a smartphone was used to create the enrollment template, then the same smartphone should be used in subsequent verification transactions in the voice recognition system.

There are also other factors which can affect the quality of voice samples other than the noise disruptions created by telephony devices. For example, factors such as mispronounced verbal phrases, different media used for enrollment and verification (using a land line telephone for the enrollment process, but then using a cell phone for the verification process), as well as the emotional and physical conditions of the individual.

Finally, the voice samples are converted from an analog format to a digital format for processing. These raw voice data types are inputted into a spectrograph, which is a pure visualization of the acoustic properties of the individual's voice. As can be seen, the two primary axes are frequency (the vertical axis) and time (the horizontal axis).

The next steps are unique feature extraction and creation of the template. The extraction algorithms look for unique patterns in the individual's voice samples. To create the template, a "model" of the voice is created. There are two statistical techniques which are primarily utilized when formulating the voice recognition biometric templates. They are as follows:

1. HMMs: This is a statistical model which is used with text-dependent systems. This type of model displays such variables as the changes and the fluctuations of the voice over a certain period of time, which is a direct function of the pitch, duration, dynamics, and the quality of the person's speaking voice;
2. Gaussian mixture models, or GMMs: This is a state mapping model, in which various types and kinds of vector states are created which represent the unique sound characteristics of the particular individual. Unlike HMMs, the GMMs are devoted exclusively for use by text-independent systems.

However, just like other biometric technologies, there are other external factors which can impact the quality of the voice recorded (this recorded voice can also be considered the "raw image") and, subsequently, affect the quality of the enrollment and verification templates created. These factors are as follows:

1. Any type or kind of misspoken, misunderstood, or misread text phrases the end user is supposed to recite;
2. The emotional state of the individual when they are reciting their specific text or phrase;
3. Poor room sound acoustics;
4. Media mismatch (for instance, using different telephones or microphones for both the enrollment and verification phases);
5. The physical state of the individual (for instance, if they are sick and suffering from any other kind or type of ailment);
6. The age of the individual (for example, the vocal tract changes as we get older in age).

## The advantages and disadvantages of voice recognition

When voice recognition is compared against the seven major criteria, it is a mixed result which comes out in the end:

1. Universality: Voice recognition is not at all language dependent. This is probably its biggest strength. As long as an individual can speak,

theoretically, they can be easily enrolled into a voice recognition system;

2. Uniqueness: Unlike some of the other biometrics, such as the iris and the retina, the voice does not possess as many unique features which results in the lack of rich information and data;
3. Permanence: Over the lifespan of an individual, and with age, the voice changes for many reasons which include age, fatigue, any type or kind of disease of the vocal cords, as well as the emotional state and any medication the individual might be under;
4. Collectability: This is probably the biggest disadvantage of voice recognition. For example, any type or kind of variability in the medium which is used to collect the raw voice sample can greatly affect or skew the voice recognition biometric templates. Therefore, it is of utmost importance to ensure that the same type and kind of collection medium is used for both the enrollment and verification stages. There should be no interchangeability involved whatsoever;
5. Performance: Because of the variableness involved when collecting the raw voice samples, it is difficult to put a gauge on how well a voice recognition system can actually perform. Also, the template size can be very large, on the range of 1,500–3,000 bytes;
6. Acceptability: This is probably one of the strongest advantages of voice recognition. The technology is actually very nonintrusive and can be deployed in a manner which is very covert to the end user;
7. Resistance to circumvention: When compared to the other biometric technologies, voice recognition, to a certain degree, can be easily spoofed, by one end user mimicking the voice acoustics of another end user. This is in large part due to the lack of unique features in the voice itself.

## The market applications of voice recognition

The market applications of voice recognition have been much more limited when compared to the other biometric technologies of hand geometry recognition, fingerprint recognition, and even iris recognition. A part of the reason for this is that there is a lack of vendors actually developing voice recognition solutions.

But, on the whole, voice recognition is starting to get some serious traction, as businesses and governments worldwide have started to realize its ease of deployment and the other strong advantages it offers to the marketplace. Probably one of the biggest market applications voice recognition is used for is in the financial world.

For example, many of the smaller to medium-sized brokerage institutions offer voice recognition to their customers as a means of quick verification.

Rather than wasting a customer's time by inputting their PIN and Social Security numbers, a customer can be identified very quickly by the use of voice recognition. What would normally take minutes to authenticate a customer with the traditional means of security is now literally shaved down to just mere seconds.

As a result, financial transactions for the customer can occur at a very quick pace. Also, voice recognition can be used on smartphones as a means of verification instead of having to enter a numerical-based password. Voice recognition is also currently being used in correctional facilities (to monitor the telephone privileges of the inmates), the railroad system, border protection and control, as well as certain types and kinds of physical and logical access entries.

This section of the book now deals with behavioral biometrics. With this in mind, the unique features to be extracted are all based upon the variations in each person's own mannerisms, upon which they can be verified and/or identified. We first start off with signature recognition.

## SIGNATURE RECOGNITION

For the longest time, signatures have been associated with law enforcement and identity, much in the same way the fingerprint has been. But unlike the other biometrics, the use of a signature actually creates a legal binding contract or agreement. As a result, the interest in the use of signature recognition has generated much more interest than just as a tool for confirming the identity of a particular individual.

In other words, the use of a signature can do both, proliferated by the legislation enacted by the Clinton Administration which makes signatures even legally binding in e-commerce transactions. The most common form of signature with which we are the most familiar with is the type when we sign our name at the POS terminal at our local stores.

For instance, when you swipe your credit card through the POS system, you are prompted to sign your name. The image of your signature then gets transmitted to the sensor pad, and from there, to the credit card database where specialized software then matches the signature which is stored in the central databases to what has just been presented.

If the match is authentic, the transaction is then subsequently approved, and the customer can then procure their products. But there is often confusion between this type of signature and a "digital signature", as well as a major difference. For example, digital signatures are not based upon the actual, physical signature itself.

Rather, it is based upon encrypted bit strings which are attached to electronic documents, which can be used to authenticate and approve all types and kinds of electronic transactions, and to legally approve and bind "paperless documents" and records.

## The differences between a signature and signature recognition

And in this same regard, there is also confusion between the actual, physical signature just described, and the signature recognition itself. With signature recognition, it is very important to note that it is not the image of the actual physical signature which is stored and compared in the signature recognition database.

Rather, it is the behavioral patterns inherent to the process of signing which is stored and compared. This includes the changes in timing, pressure, and speed in which the signature is composed. Although it is comparatively straightforward to duplicate the visual appearance of a signature, it is very difficult to duplicate behavioral characteristics. This process is known as "dynamic signature recognition".

## Signature recognition – how it works

It should be noted at this point that signature recognition is primarily used for verification purposes only. While in the most theoretical sense it could potentially be used for identification applications, in the real world, it would prove to be very impractical.

This is so for a number of reasons. First, the hand can be greatly affected by genetic factors, and various types and kinds of physical features (such as ailments, and the aging process of the hand we all experience at some point in time or another). Second, the signature is very dynamic and can change very quickly over time, whether the individual has intentions to or not in altering their own signature. Third, unlike the other biometric technologies, there is virtually no permanence or long-term stability (unlike the iris or the retina) associated with signature recognition, because of its ever-changing nature.

The very first signature recognition devices utilized static variables such as the height, spacing, slope, as well as the various characteristics in terms of the shaping of the letters which are found in the signature. By the mid-1970s, signature recognition became much more dynamic in the sense that various spatial, pressure, and temporal variables were taken into consideration. These variables included such factors as the downward pressure which was applied to the pen, the level of pressure at which the pen itself is gripped at, and the angle at which the pen is held while the individual signs his or her name, and even the time it takes for the signature to be completed.

Today's signature recognition devices can now collect and analyze such variables as speed, acceleration, pauses, and the changes in pressure in which the individual signs their name on the special writing tablet. Neural network technology can also be incorporated with signature recognition which can literally learn the ever so slightest changes and variations in the way an individual signs their name over a preestablished period of time and make the necessary changes to the database.

Signature recognition technology involves the use of a pen and a special writing tablet, which are connected to a local or central computer for processing and verification. To acquire the signature data during the enrollment process, an individual is required to sign his or her name several times on the writing tablet. It should be noted that the robustness of the signature recognition enrollment template is a direct function of the quality of the writing tablet.

A high-quality tablet will capture all of the behavioral variables (timing, pressure, and speed). In contrast, a low-spec tablet may not be able to capture all these variables. There are several constraints to the data acquisition phase. First, a signature cannot be too long or too short. If a signature is too long, too much behavioral data will be presented. As a result, it will be difficult for the signature recognition system to identify consistent and unique data points.

If a signature is too short, insufficient data will be captured, giving rise to a higher false accept rate. Second, the individual must complete the overall enrollment and verification process in the same type of environment and under the same conditions.

For example, if the individual stands during enrollment but sits down during verification, the enrollment and verification templates may vary substantially (this is attributable to the amount of support given to the arm). Once the data acquisition phase has been completed, the signature recognition system extracts unique features from the behavioral characteristics, which includes the time needed to place a signature, the pressure applied by the pen to the writing tablet, the speed with which the signature is placed, the overall size of the signature, and the quantity as well as direction of the signature strokes.

With signature recognition templates, different values or "weights" are assigned to each unique feature. These templates are therefore as small as 3 kb. One of the biggest challenges in signature recognition is the constant variability in the signatures themselves. This is primarily due to the fact that an individual never signs their signature in the same fashion any given two successive times.

For example, the writing slope can switch tangentially left to right (and vice versa), and up and down (and also vice versa); the exact pressure put on the pen can change greatly each and every time the individual has to submit a verification template; and even a reflective light on the surface of the signature recognition capture device can indirectly cause variances in the speed as well as the timing of the signature.

## Signature recognition – the advantages and the disadvantages

The most significant benefit of signature recognition is that it is highly resistant to impostors. For example, while it is quite easy to forge a signature, it is very difficult to "mimic" the behavioral patterns inherent to the process of signing. Signature recognition is well suited to high-value transactions. For example, signature recognition could be used to positively verify the business representatives involved in a transaction before any classified documents are opened and signed.

Second, compared to the other biometric technologies, signature recognition is deemed to be non-invasive. As a result, the technology could be widely accepted by users. We have all used our signature to authorize transactions, which does not in any way impede on our sense of privacy (or privacy-related rights).

Third, there is always a concern among users of biometric systems that templates may be stolen if the system is compromised, and that stolen templates cannot be replaced and/or changed. This may be true for physiological biometrics (it would, for example, be very difficult to change the structure of your fingerprint or iris). Signature recognition allows the behavioral dynamics of the way you place your signature to be changed, thus easing user concerns.

On the downside, signature recognition is prone to higher error rates, particularly when the behavioral characteristics of signatures are mutually inconsistent. In addition, users may have difficulties getting used to signature tablets, which could also result in higher error rates.

Signature recognition can also be compared against the seven criteria, with the end results being that the number of advantages it offers balances out for the most part, with its disadvantages:

1. Universality: Probably its biggest strength, signature recognition can be used in almost any type and kind of language, ranging all the way from English to the most obscure languages ever heard of;
2. Uniqueness: Obviously, the more characteristics a signature has, the more unique it becomes. As mentioned, when the variables of speed, velocity, timing, and pressure are introduced, dynamic signatures can virtually never be lost or stolen;
3. Permanence: This is probably one of the greatest weaknesses of signature recognition. An individual's signature can vary greatly over just a span of a few minutes, by such variables as fatigue, illness, stress, or distraction. These variables, in turn, can also affect the ability of the individual to actually grip the writing pen used to create the dynamic signature and exert the downward pressure which is needed;
4. Collectability: The ability to collect raw signature samples is a direct correlation of the quality of the signature recognition system itself

(meaning, a lower grade signature system will make it much more difficult to collect a robust signature sample);

5. Performance: It has been recognized that the best performing signature recognition devices are about 96% accurate. But the better the signature recognition device is, the higher the cost will be. A strong advantage here is that signature recognition requires no end user training;
6. Acceptability: In general, when compared to the other biometric technologies, most people are very comfortable with providing their signature, primarily because of its ease of use and non-privacy rights issues. It is also important to note that the theft of personal data is much less likely with dynamic recognition technology versus graphic recognition technology (such as storing a fingerprint or an iris);
7. Resistance to circumvention: This is one of signature recognition's other biggest advantages: A signature which is dynamically produced is almost impossible to forge.

In summary, out of all biometric technologies, whether physical or behavioral, signature recognition offers most potential in terms of adaptability and implementation. This holds true from a number of perspectives.

First, there is its ease of use – the user simply places his/her signature as he/she would normally do.

There is, therefore, no need for end user training, as is the case with other types of biometric technology, including facial recognition and retinal recognition. The training needed for the person operating the signature recognition device is also minimal.

Second, the implementation cost of signature recognition is low – the system consists of a special pen, a tablet, and software (which, in the case of retailers, could be installed in a POS terminal). Costs are minimal compared to those of much more complicated systems, such as a retinal scanning device.

Third, a signature recognition system can easily be embedded in an organization's prevailing security processes, without excessively disrupting or affecting existing operations. To give an example, no major wiring or installation is needed (as is the case with a hand geometry scanner or fingerprint scanner). Signature recognition could therefore prove a highly valuable tool for multimodal security solutions.

## KEYSTROKE RECOGNITION

Keystroke recognition is the next behavioral biometric technology to be examined. As its name implies, keystroke recognition relies upon the unique way in which we type onto a computer keyboard. In today's mobile world and global society, typing is something which we take for granted every day.

Whether it is the way we type onto our keyboard on our office or home computer, or even our smartphone, there is a distinct way in which we type. This uniqueness comes from most specifically in the rhythm in which we type, for how long we hold the keys down on our smartphone or computer keyboard, and the succession of keys used overall in the typing process.

In this absolute regard, keystroke recognition can actually be considered to be the oldest biometric technology around, even much more so than hand geometry recognition or fingerprint recognition. This is primarily so because the interest in unique typing patterns dates all the way back to the 19th century, when the Morse code first came out.

By World War II, the US military intelligence department could actually identify enemy Morse code operators by their unique typing patterns. Although the Morse code is only, technically speaking, a series of dots and dashes, some distinctiveness could still be established.

The first keystroke recognition device came out in 1979, and by 1980, the National Science Foundation scientifically validated the technology of keystroke recognition, and by 2000, it was finally accepted as a commercial biometric technology, which could be used in either the public or private sector.

## Keystroke recognition: how it works

To start the enrollment process, an individual is required to type a specific word or group of words (text or phrases). In most cases, the individual's user name and password are used. It is very important that the same words or phrases are used during both the enrollment and verification processes. If not, the behavioral typing characteristics will be significantly different, and, as a result, a mismatch will arise between the enrollment and verification templates.

To create the enrollment template, the individual must type his or her user name and password (or text/phrase) about 15 times. Ideally, the enrollment process covers a period of time, rather than taking place all at once. This way, the capture of behavioral characteristics will be more consistent.

With keystroke recognition, the individual being enrolled should type without making any corrections (for example, using the backspace or delete key to correct any mistakes). If the individual does make corrections, the keystroke recognition system will prompt the individual to start again from scratch. The distinctive, behavioral characteristics measured by keystroke recognition include the following:

1. The cumulative typing speed;
2. The time that elapses between consecutive keystrokes;
3. The time that each key is held down (also known as the dwell time);

4. The frequency with which other keys, such as the number pad or function keys, are used;
5. The key release and timing in the sequence used to type a capital letter (whether the shift or letter key is released first);
6. The length of time it takes an individual to move from one key to another (also known as the flight time);
7. Any error rates, such as using the backspace key.

These behavioral characteristics are subsequently used to create statistical profiles, which essentially serve as enrollment and verification templates. The templates also store the actual user name and password. The statistical profiles can either be "global" or "local". Whereas a "global" profile combines all behavioral characteristics, a local profile measures the behavioral characteristics for each keystroke.

The statistical correlation between the enrollment and verification templates can subsequently be modified, depending on the desired security level. An application which requires a lower level of security will permit some differences in typing behavior. However, an application which requires a higher level of security will not permit any behavioral differences.

It is important at this point to make a distinction between static and dynamic keystroke verification. In case of the former, verification takes place only at certain times – when the individual logs in to his or her computer, for example. With the latter, the individual's keystroke and typing patterns are recorded for the duration of a given session.

## Keystroke recognition: the advantages and the disadvantages

Keystroke recognition has several strengths and weaknesses. Arguably its biggest strength is that it doesn't require any additional, specialized hardware. As previously indicated, keystroke recognition is purely software-based, allowing the system to be set up very quickly.

Second, keystroke recognition can be easily integrated with other, existing authentication processes. The adoption of other biometric technologies requires the implementation of a new process within an existing process. This calls for individuals who are properly trained in the use of contemporary biometric devices, which can greatly increase costs.

Third, everybody is familiar with typing their user name and password. As a result, there is very little training required for an individual to use a keystroke recognition system properly. Fourth, the templates that are generated by the system are specific only to the user name and password used. Should this user name and/or password be tampered with, the individual only needs to select a new user name and password to create a new set of enrollment and verification templates.

The weaknesses of a keystroke recognition system are the same as those suffered by other systems that rely on a user name/password combination. For example, passwords can be forgotten or compromised while users will have to remember multiple passwords in order to gain access to, for example, a corporate network. It should be noted that keystroke recognition still requires users to remember multiple passwords (the administrative costs of having to reset passwords will also continue to be incurred).

As such, it only enhances the security of an existing user name/password-based system. Second, keystroke recognition is not yet a proven technology. As a result, it has not been widely tested. And finally, keystroke recognition is not necessarily a convenient system to use.

Just like signature recognition, keystroke recognition is not as widely implemented as some of the other biometric technologies, such as fingerprint recognition and iris recognition. It too can be evaluated against the seven criteria:

1. Universality: This is a key strength of keystroke recognition, even people who are "one-finger" typists or not even familiar with typing at all can still be accommodated;
2. Uniqueness: At the present time, keystroke recognition only possesses enough unique features to be used for verification applications and not for identity applications;
3. Permanence: In this regard, this is one of keystroke recognition's biggest weaknesses, as the typing pattern of an individual can change due to injuries, disease, increased typing proficiency, fatigue, lack of attention, or even using a different keyboard can cause an individual to have a different typing pattern and rhythm;
4. Collectability: It can take many typing samples until enough unique features can be extracted;
5. Performance: When the proper security threshold setting is established by the systems administrator, keystroke recognition can produce an FRR of up to 3%, and an FAR of up to 0.01%. As mentioned, keystroke recognition does not require any additional hardware, enrollment and verification can happen remotely, and it can even be used to further security harden passwords, and the template size is quite small;
6. Acceptability: There are no privacy rights issues with keystroke recognition, and there are no negative correlations associated with typing;
7. Resistance to circumvention: Any typing data which is not encrypted can be used maliciously by a third party and even be used to spoof the keystroke recognition system. Also, key loggers can be established onto the computer itself to record the various keystroke patterns and rhythms.

Keystroke recognition is included in very few commercial applications at this point. Compared with other biometric technologies, it probably ranks last in terms of use. Having said that, behavioral biometrics are not particularly popular on the whole. Although it's quite possible that keystroke recognition will establish itself as a dominant technology, there are, at this stage, not enough vendors to propel the technology forward (unlike, for example, fingerprint recognition).

Compared with other physical biometrics, keystroke recognition is easier and cheaper to implement. However, it is unlikely to be used for applications such as physical access control, document verification, and passport verification. Instead, it will be used for computer security (where fingerprint and iris recognition solutions are already used as a substitute for user names and passwords).

Keystroke recognition is also well suited to e-commerce applications. Here, a user would be able to access an Internet banking or e-commerce site by typing in the same text or phrase several times (rather than having to remember different user names and passwords). Moreover, the same text or phrase can be used to log into multiple e-commerce sites. Keystroke recognition could additionally be the security tool of choice for multi modal security applications, where it can be used to provide third, fourth, or even fifth tier security.

While small to medium-sized enterprises (SMEs) will probably not adopt keystroke recognition, it is well suited to large businesses and organizations, including major banks and financial institutions. It's also quite conceivable that keystroke recognition will be adopted by governments around the world.

## BIOMETRIC TECHNOLOGIES OF THE FUTURE

In summary, thus far, we have reviewed all of the physical-based and behavioral biometrics, with some of the biometric technologies being seen in more commercial usage than the others. But there is always continuous research and development always going on in biometrics, in order to see what the future holds. The following are the biometric technologies which show the greatest promise thus far in terms of development and potential use:

1. DNA Recognition;
2. Earlobe Recognition;
3. Gait Recognition.

The use of DNA as a means of being used as a unique piece of evidence did not start until 1984, by a scientist known as Alec Jeffreys. It did not make its mark in the United States until at least the late 1980s, when the FBI invested much resources in exploring the use of DNA as a means of criminal evidence which could be presented in a court of law.

As a result of this massive effort, one of the world's largest databases was born. This is today known as the Combined DNA Index System, or CODIS for short. This database is now used by law enforcement agencies at all levels throughout the United States, as well as at the international level. By 2012, this database held the DNA records of more than 1.2 million individuals, primarily those of criminals, suspects, and illegal immigrants.

At the present time, DNA is only used for forensic purposes, often collected as a means of latent evidence. It can only be used at the present time for identification purposes, and not for verification applications, due to the time that it takes to process the raw data which is collected, which is about 4 to 5 hours in total length to complete.

But due to the uniqueness and the richness of the data that DNA possesses, there is thus a strong interest in the use of it as a potential biometric technology. The use of DNA makes strict use of the patterns which are found in deoxyribonucleic acid and is located in the cell nucleus of all living beings, no matter how large or how small.

## DNA RECOGNITION: HOW IT WORKS

DNA profiling makes use of the common, repetitive patterns of the four base pairs, which are adenine (A), cytosine (C), guanine (G), and thymine (T). It should be noted that DNA profiling occurs in the uncoded portion of the DNA sequence, which is simply known as the uncoded DNA, for which there is no identified function. A base pair pattern can look like this: AGCCTTCAGTA.

This sequence is also known as a short tandem repeat, or an STR for short. A unique STR exists when two or more base pairs appear in a certain and repetitive pattern, and also when this exact pattern is repeated close by in adjacent spots. It should be noted that the STR is typically anywhere from two to four nucleotide base letters, and even larger patterns, known as variable number tandem repeats, or VNTR for short, can contain up to 80 letters. This thus provides for a much higher degree of accuracy and probability. But the analysis can take up to days to complete, not just a matter of a few short hours.

There are two kinds of DNA which are currently present in all living beings, and they are as follows:

1. Nuclear DNA: This is the contribution of genetic components from both the mother and the father and is most commonly found in the blood, saliva, semen, and even the bones;
2. Mitochondrial DNA: This is the contribution of genetic components from the mother only and can also be found in the hair and the teeth. This is the type of DNA which is used to identify missing individuals.

The raw DNA sample can be collected from blood, semen stains, saliva, hair, or even under voluntary conditions. With the latter approach, a buccal swab can be very easily inserted into the mouth, in order to collect a sample of DNA. After the DNA sample is collected from the specific source, the DNA is then isolated and then further divided into much shorter segments (also known as restriction enzymes) which terminate the DNA strands at various points, and differentiated in size using an electric current.

The differentiated sizes are then placed onto a nitrocellulose filter which possesses different fluorescent dyes, which then further attach themselves to the differing repeating patterns. This is then X-rayed, and the resultant image is a DNA fingerprint.

Subsequently, this DNA fingerprint is then translated into a DNA profile which displays the number of STRs at the specific locations. The DNA markers used to construct this DNA profile are located on chromosome numbers 1, 2, 4, 5, 10, and 17. These specific locations are referred to as DNA loci. After this profile is constructed, the findings occur in one of the following three categories:

1. Exclusion: This result possesses the highest level of certainty (meaning, the raw data is different from the source it was collected from);
2. Non-exclusion: The raw data cannot be excluded of coming from the same source (in other words, the raw DNA sample and the source from which it was collected from are the same);
3. No result: No conclusions can be drawn.

## DNA recognition: the advantages and the disadvantages

Because DNA recognition is still under heavy research and development, it can only be matched up against five of the seven criteria, which are as follows:

1. Uniqueness: Among all of the other biometrics, DNA possesses the most unique and richest information and data (even much more so than the retina), and if it proves to be a viable biometric technology, it will far surpass the levels of the other biometrics which currently exist in the marketplace;
2. Collectability: The right quality and quantity of DNA must be present, and the material must not be contaminated;
3. Performance: DNA analysis is very useful and robust for forensic types of applications, but for verification purposes, it is still not useful, because the 4–5 hours of analysis which is required at present must be brought down to a matter of mere seconds;

4. Acceptability: DNA is very much prone to privacy rights issues, and while the DNA profile itself does possess genetic information about an individual, the raw DNA data does have this;
5. Resistance to circumvention: Although DNA is rich in terms of data and information, it can often be very easily substituted for another person's DNA.

Other disadvantages of DNA recognition include the following:

1. DNA samples are prone to degradation and contaminants;
2. Implementing the security protocols for a DNA-based biometric system could prove exceptionally complex. This applies to database protection and overall system confidentiality in equal measure.

## GAIT RECOGNITION

Gait recognition is the next potential biometric which is being looked at very closely as a viable technology into the future. Gait recognition, simply put, is the potential verification and/or identification of an individual based upon their unique stride, or the unique way in which one walks in normal, everyday life. It can be considered both a behavioral- and physical-based potential biometric and thus is considered to be part of the field known as kinematics, which is the study of motion.

With our stride, the various aspects of the body closely position themselves at unique lengths, angles, and speeds. At the present time, there is strong interest in using gait recognition for identification purposes, because the raw data which is needed to create the biometric templates can be collected in a very covert fashion. As a result, gait recognition can be used very well with CCTV technology and facial recognition.

### The process behind gait recognition

There are currently two methods which are being investigated for capturing and examining the unique strides in individuals:

1. Static shapes: This methodology actually takes a silhouette pattern of a walking person from a series of vide of video frames taken by a CCTV camera. This image of the walking individual is defined in a boxed region of the video frame, and any extraneous features from the external environment are removed. The silhouette image is then created, and from there, a further series of images are created, then digitally mapped to yield what is known as the gait signature;

2. Dynamic shapes: This type of methodology uses an accelerometer, placed on the individual's leg, and the stride (walking) motion is then collected and analyzed from three different directions.

There are numerous methods which are currently being investigated to capture the raw data, and they are as follows:

1. Machine vision: With this, the individual video frames which display the individual walking are isolated from one another frame by frame;
2. Doppler radar: With this system, a sound wave is bounced off of the leg of the individual, and the return sound wave is then measured. This is done by ascertaining the velocity of the signal which is modulated by the moving parts of the leg;
3. Thermography: This technique is embedded into a computer vision system in order to cast out the moving leg in difficult and extraneous lighting situations.

After the raw images are captured and collected, the next step in the biometric process is unique feature extraction. Currently, there are two methodologies which are also being examined to accomplish this task:

1. Model-based analysis: With this technique, the height of the individual, the distance between the head and the pelvis, the total distance between the pelvis and the feet, as well as the total distances between the moving feet are measured and calculated. Also, the relationships of the distances and the angles of the walking individual can be converted into various vectors as well;
2. Holistic analysis: With this technique, the silhouette of the walking individual is also calculated and measured.

Finally, after unique feature extraction, the next step is the creation of the biometric templates. In this regard, gait recognition is different from the other biometric technologies in that actual sequences of movements are compared against one another. There are also various template creation techniques which are under research and development, and they are as follows:

1. Direct matching: With this approach, a gait sequence is identified by comparing the raw data which is captured with the reference data, and from there, the various distances which yield the least minimal distances are subsequently calculated;
2. Dynamic time warping: With this technique, the closeness between two sequences of gait movements is measured. The variables here

examined and analyzed include the time, speed, and the various types and kinds of walking patterns;

3. HMMs: This is a statistical-based methodology in which the probability of the similarity of the shapes which appear in the walking stride of an individual is examined and analyzed in succession.

## Gait recognition: the advantages and the disadvantages

Gait recognition, although still under research and development, can be compared and evaluated against the same seven criteria the other biometric technologies have been rated against:

1. Universality: The collection of raw data requires that an individual possess a walking stride which is basically unimpeded and be without the use of such walking aids such as crutches, walkers, and even wheelchairs;
2. Uniqueness: Since gait recognition is not yet a viable biometric technology, it's true, absolute uniqueness still cannot yet be ascertained when compared to the other biometric technologies, and as a result, at the present time, gait recognition is predicted to work very well in a potential multimodal biometric security solution;
3. Permanence: A person's walking stride can be affected by such variables as older age, physical injuries, or even disease. The walking stride can also be affected by such variables as the type of surface the individual is walking upon, any additional or extraneous weight which is being carried by the individual, and even the type and kind of footwear the individual uses while they are walking;
4. Collectability: A potential advantage of gait recognition is that raw data which has a low resolution as captured by the CCTV system can still be used to create robust biometric templates;
5. Performance: Some of the studies conducted thus far have claimed that gait recognition possesses an ERR of 5% (and perhaps even less);
6. Acceptability: It appears that gait recognition can be considered to be potentially a non-invasive biometric, but the issues of privacy rights and civil liberties violations could emerge as a person's unique walking stride can be collected and analyzed covertly;
7. Resistance to circumvention: At this point in research and development, it is deemed that it could be very difficult to mimic another individual's walking stride.

Finally, it is forecasted that gait recognition can be used as a great surveillance tool (especially when combined with CCTV technology and facial recognition) in large venue settings, such as sporting events, concerts, transportation hubs, and educational institutions. Also, gait recognition can be used to track suspicious behaviors with regard to motion of an individual, for example, if a particular person appears in the CCTV field of view an unusual number of times in a short time frame.

## EARLOBE RECOGNITION

Earlobe recognition is the last potential biometric technology to be examined in this chapter. Its studies for the examination of its unique features date all the way back to the 19th century, and an ear classification system was first proposed way back in 1964 by a scientist known as Alfred Iannerreli. In the mid-2000s, the European Commission further studied the use of the ear as a potential biometric.

Earlobe recognition is a potential physical biometric, in which the unique features are extracted and examined. The ear is fully shaped by the time an infant turns four months of age. As a potential biometric, it is the structure of the outer ear which is of prime interest.

### Earlobe recognition: how it works

Current research is focused upon using a technique known as "Eigenears", which is based upon the different structures of the outer ear. Eigenears is very similar to the technique of "Eigenfaces" used in facial recognition. Thus far, there are positive results in the scientific community that the outer ear is unique in its overall shape, in terms of depth, angular structure, and the overall formation of the earlobe structure.

From within this, scientists are currently examining the distances between the regions of the ear, in 45-degree and three-dimensional angular models. The raw data of the ear can be captured by taking an actual picture of the outer ear, preferably when the ear is placed against a glass platen in order to capture the actual, raw images. These raw images are then used for unique feature extraction and biometric template creation. Either a three-dimensional or a thermal picture can be taken of the outer ear, with preference being given to the latter.

At the present time, there are two methodologies which are currently being examined for potential earlobe recognition, and they are as follows:

1. PCA: With this technique, the raw image of the ear is cropped, it is then scaled to a regular size, thus capturing the two key points of the ear, specifically the regions known as the triangular fossa and the antitragus, from which the unique features can be extracted;
2. Three-dimensional analysis: With this technique, the outer image of the ear is captured, thus ascertaining both the depth and the color type of the ear. Both the helix and the antihelix are examined, from which the unique features can be captured.

## Earlobe recognition: the advantages and the disadvantages

Also, the potential of earlobe recognition can also be examined from the same seven criteria that all of the other biometric technologies in this chapter have been analyzed against:

1. Universality: For the most part, unless an individual has a physical injury which caused them to have an absence of the ear, most people have an ear which can be examined for unique features;
2. Uniqueness: Although it is still not yet proven scientifically, preliminary results have shown that the two tests involving 10,000 earlobes and identical twins, all individuals were to have found to each to possess a unique ear structure;
3. Permanence: The shape of the ear more or less stays the same over the lifetime of an individual;
4. Collectability: Certain variables from the external environment can certainly impact the capture of the raw images of the ear, and these variables include different types and kinds of lighting situations, jewelry such as the wearing of earrings, and wireless device accessories which can be placed in the outer ear;
5. Performance: It is still too early to determine how well earlobe recognition will actually be adopted and used in the commercial market;
6. Acceptability: Earlobe recognition is deemed to be a potentially nonintrusive type of potential biometric technology; thus, there should be no issues with regard to privacy rights;
7. Resistance to circumvention: At this point, it is difficult to determine this, but it is possible to construct a false earlobe.

Since there are no commercial applications yet of earlobe recognition, in the future, it is heavily anticipated that it will work well with facial recognition systems, in multimodal biometric solution applications.

Finally, of the three potential biometric technologies covered in this chapter, gait recognition appears to offer most potential in terms of practical deployment. Its potential is very strong for multi-modal security applications. Think, for example, of a dual-level verification system whereby a subject's identity is initially established on the basis of, for example, a fingerprint before being verified using gait recognition. It is also useful for one-to-many verification requirements at, for example, airports.

Under these conditions, gait recognition can offer significant time savings by allowing large groups of people to be identified in a single environment. Compared to gait recognition, earlobe recognition has much further to go, also in terms of research and development. This also applies to DNA recognition, even though it has the potential to be the best biometric of all (on account of its uniqueness).

## REVIEW

In review, this chapter examined the biometric technologies which are in use today, and what is being planned for the future. More specifically, the physical biometric technologies were examined (which include hand, finger, retinal, iris, vein pattern, palm print, and facial recognition), as well as the behavioral biometrics were also reviewed (which include keystroke and signature recognition). Also, the biometric technologies of the future were also looked at (which include DNA, earlobe, and gait recognition).

Out of all of these biometric technologies, it is primarily fingerprint, iris, and vein pattern recognition which are currently being deployed the most at the present time. For the future, it is gait recognition which holds the most promise of being a viable biometric technology. This fact brings up a very important subject which is very important to any CIO as well as any other C-Level Executive. And that is the overall topic of biometric system analysis and design.

For example, before any biometric technology can be considered for financial acquisition and ultimate deployment, the existing security infrastructure of the place of business or organization must be very carefully mapped and examined in order to fully determine just how well a particular biometric technology can fit into.

For instance, key questions must be carefully evaluated and answered. Some of these questions include the following:

1. Will the particular biometric technology being considered be deployed as a standalone system (in other words, will it be the primary means of defense for the business or organization)?
2. Or will it be implemented as a security add on within the existing security infrastructure (such as a multimodal security solution, where there are multiple layers of security)?

Will the biometric technology being considered for deployment be wired and networked within the existing security infrastructure? Or will a different networking environment be needed?

## Notes

1 Certified Biometrics Learning System, Module 2, Biometrics Standards, Copyright 2010 IEEE, pp. 2–1.
2 Certified Biometrics Learning System, Module 2, Biometrics Standards, Copyright 2010 IEEE, pp. 2–1
3 "Retina Identification", Robert "Buzz" Hill. Article is from the book: *Biometrics: Personal Identification in Networked Society*, by Anil Jain, Ruud Bolle, and Sharath Pankanti, p. 124.

Chapter 3

# Biometric project management

The goal of this chapter is to explore, examine, and provide answers to these particular questions and much more. This chapter is intended to be an introductory project tool for the C-Level Executive, as they contemplate the choice of biometric technology which be the most optimal and work best for their place of business or organization.

Specifically, this chapter is divided into the following topics:

1. Biometric technology system architecture;
2. Biometric system analysis and design processes (the biometrics project management section)
3. Biometric networking topologies.

Of course, this chapter is not intended to be a substitute for professional project management advice. Rather, the goal is to give the C-Level Executive the starting point they need in which to determine the serious questions which need to get asked (as well answered) in order to get the maximum use of their ever-tightening security budget.

## BIOMETRIC TECHNOLOGY SYSTEM ARCHITECTURE

Generally speaking, biometric devices, no matter how large or small they might be, are actually quite complex, even though they may also look very simple in nature on the outside in terms of design and ergonomics. Of course, though, the exact technology components will vary from vendor to vendor, but all biometric technologies (even whether it is physical or behavioral-based) consist of the following subcomponents:

1. Sensing and data acquisition;
2. Signal and image processing;
3. Data storage;
4. Template matching;


DOI: 10.4324/9781003562146-3

5. Threshold decision-making;
6. Administration decision-making;
7. Data (biometric template) transmission.

## SENSING AND DATA ACQUISITION

Each of these subcomponents will be reviewed in much further detail in this chapter, first starting with the sensing and data subcomponent. In its simplest terms, this subcomponent involves the capture and the extraction of raw biometric data (in particular the image), and from that point onwards, converting that raw data (the image) into a form which can be used for further processing from within the particular biometric technology.

This subcomponent can collect two types of raw data, and this happens at first in the enrollment stage, and then at the secondary verification and/or identification stage. These two types of raw data which can be collected are defined as follows:

1. Probe data: This is the first series of raw data to be collected, and it is used to create the enrollment template;
2. Live sample data: This is the second series of raw data to be captured, and this is used to create the subsequent verification templates (which in turn is then used to compare against the enrollment template for various types and kinds of verification and/or identification applications).

Since this subcomponent primarily deals with the collection of biometric raw data, obviously the way, manner, and technique in which either the physical or behavioral biometric is presented to the sensing device thus becomes of prime importance. Specifically, in this regard, the term presentation can defined as "the submission of a single biometric sample on part of the user". And, the factors which affect presentation can be specifically defined as "the broad category of variables affecting the way in which the users' inherent biometric characteristics are displayed to the sensor" (Source: Certified Biometrics Learning System, Module 2, Biometrics System Design & Performance, Copyright 2010 IEEE, pp. 3–4).

Therefore, the standardization process of presentation also becomes of prime importance, and involves ensuring the following in order to establish a consistent, raw sample acquisition process:

1. The ease of use in ergonomic sensor design;
2. Effective user and operator/administrator training (extreme familiarity with the biometric system is an absolute must here);

3. Quality feedback mechanisms are put into place on the part of both the end user and administrator so that the presentation technique can be constantly improved upon.

## MULTIMODAL BIOMETRIC SYSTEMS

In today's biometric deployments, rarely do you find just one device being used. Rather, you see multiple devices being used, thus requiring multiple sensing devices. In this regard, the following sensor schemes can be established:

1. One sensor of multiple devices of the same biometric modality;
2. Multiple sensors of multiple devices of differing biometric modalities.

Such designs and layouts are known as "multimodal biometric solutions". In this regard, there are two different types of biometric multimodal systems:

1. Synchronous systems;
2. Asynchronous systems.

"Synchronous" means that there are two or possibly more biometric systems being used at the same time, for the same authorization process of the subject in question. In this scenario, imagine a fingerprint scanner and an iris scanner being used simultaneously. "Asynchronous" means that there are two (again possibly even more) biometric systems that are being used, but the key difference here is that they are used in tandem with each other, or in a sequential fashion. Again imagine the fingerprint scanner and the iris scanner. The fingerprint scanner would be used first and immediately after, the iris would be scanned to confirm the identity of the subject in question.

There are two key points to remember when using biometric multimodal systems:

1. Using multiple systems increases the accuracy, although it does not increase the speed of the authorization process.
2. If one device does not work, the other systems will come into play as a fail safe, thus still assuring a strong level of security.

With regards to physical access entry, it is typically businesses and organizations that use multimodal biometric solutions. Very often, there will be a strong presence of security at the main points of access and entry of a building, and within the office infrastructure itself other rooms and private areas will be protected also. At the main points of entry, a hand geometry scanner is used to verify and confirm the identity of individuals.

*Figure 3.1* This is an image of a CCTV camera. (A close up of a metal object on a wooden post photo – free camera wallpaper image on Unsplash.)

Along with this scanner, there could be other levels of security used in conjunction, such as a swipe card or smart card system. When the individual gains entry through the main point of access, the internal parts of the office remain protected, such as the client file room and the server room. Very often, in this extra layer of security, fingerprint recognition is used. This then formulates the multimodal biometric solution in this example, and as you can see, it is an asynchronous approach. Typically, this combination of security is a very powerful one. One of the most powerful forms of a multimodal security solution is combining CCTV technology with facial recognition, see Figure 3.1.

Also, other types of biometric devices can be used in lieu of fingerprint recognition, such as facial and iris recognition. Nowadays, vein pattern recognition is starting to be used extensively as a component of a multimodal biometric package.

## Single sign-on solutions

In terms of logical access entry, passwords have long been the traditional means of security when logging onto a computer or a corporate network. But as has become clear with the proliferation of identity theft threats, passwords can be cracked and hacked very easily.

Thus, organizations and businesses are constantly faced with enforcing stronger password policies, frustrating the end user even more. But even here, biometrics is emerging as a much-favored security solution over the password. Nowadays, you can log into your network or computer with a single swipe of a finger or even a picture of your iris.

Today, new laptops often contain a small fingerprint sensor on the keyboard. While durable, significant damage can occur to it if the laptop is dropped, or exposed to harsh environmental conditions. To counter this, a secondary biometric system can be used, such as iris recognition. For example, when logging on to their laptop, an individual's identity could first be confirmed with a quick swipe of the finger, and then for extra protection, their iris could be scanned and verified via a built-in camera, before full access is granted.

As is the case with physical access entry applications, vein pattern recognition is emerging as the top choice in multimodal biometric solutions. In the world of logical access, the cardinal rule of having two or more means of identity verification is just as important, if not more.

## IMPLEMENTING A MULTIMODAL BIOMETRIC SYSTEM

In the past, implementing a multimodal biometric system was an expensive endeavor, as a biometric system would have to be implemented on top of, or with, an existing legacy security system. This would very often mean a major overhaul in the organizational infrastructure. However, the past decade has seen an evolution in biometrics technology as well as in multimodal systems.

One unit or device can now house different technologies. Nowadays, multimodal biometric solutions are not just available as hardware but can be obtained as a software development kit (SDK). This approach has become known as a "hybrid" biometric system, where fingerprint, iris, and vein pattern recognition are primarily used.

Before you are going to implement either a hardware or a software multimodal biometric system, there are some key questions to be taken into consideration:

1. As a business owner, are you trying to protect the physical or the logical aspects of your business?
2. Which biometric system will work best for your particular multimodal system?
3. Will the new multimodal biometric system be layered with the existing security infrastructure, or will it be used as an addition?
4. What are the costs, and what is your budget?
5. What kind of software applications, if any, will need to be developed to support a new multimodal biometric system?

After evaluating some of these major factors, it will soon become apparent that a multimodal biometric solution offers a number of strategic benefits. Some of these include:

1. A fortified level of security. Rather than having just one layer of security, you will now have multiple layers. Not only will there be a fail-safe mechanism in place, but should a perpetrator break through one line of defense, they will most likely be caught deeper in. In other words, an SME can retain a very high-security threshold with this scheme;
2. There are many metrics that are included in evaluating the effectiveness of a biometric system, for example, the False Accept Rate (FAR), the False Reject Rate (FRR), and the Equal Error Rate (ERR). By using a multimodal biometric system, the reliability of the verification process is greatly increased;
3. Some systems experience difficulties enrolling and verifying people of different ethnic origins, but a multimodal biometric system can capture the unique characteristics of a much larger and varied population. A read rate of almost 100% can be guaranteed;
4. In biometrics, there is always the theoretical issue that a system can be spoofed. By having extra layers of biometrics, this issue has become, over time, much less prevalent;
5. As mentioned previously, a newer version of the multimodal system is known as the hybrid approach. This permits more than one biometric technology to be used within one component. An example of this is an SDK, which allows for the different recognition technologies to be all used together, independent of the actual hardware being utilized.

## CHALLENGES WITH MULTIMODAL BIOMETRIC SYSTEMS

One of the biggest challenges to the sensing subcomponent, especially in multimodal biometric systems, is that of the physiological or behavioral trait that is presented to the biometric sensor, and the raw image which is subsequently extracted, changes each and every time. In other words, there is no such thing as two, 100% identical biometric templates that are created in the end.

There are always changes and variability that occur, based upon a number of key factors, which include:

1. The actual physiological or behavioral trait has been changed over a period of time (such as due to injuries and aging);
2. The way in which the physiological or behavioral trait is presented to the biometric sensor changes over time on the part of the end user;

3. The biometric sensor degrades over a period of time as well, due to everyday wear and tear, or even technological breakdown.

So, as one can see, the way and manner in which the physiological and/or behavioral trait is presented to the biometric sensor on the part of the end user has a direct impact on biometric system performance, and the quality of the enrollment and verification templates that are subsequently created.

The following variables are deemed to be the most important in the sensing/data acquisition subsystem phase:

1. Collectability: If the physiological or behavioral trait cannot be measured by the biometric system, no matter how good the particular trait may be, it will be deemed to be useless;
2. Performance: The quality of both the biometric sensor and the raw image must be to or even exceed the required performance levels;
3. Acceptability: The method of sensing must prove to be very easy to use and hazard-free by the end user;
4. Circumvention: The biometric sensor must be 100% "spoof-proof".

A function that is very important in presenting a physiological or behavioral-based biometric trait to the biometric sensor is that of the attempts which are made by the end user. Specifically, as it relates to biometrics, an attempt can be described as: "the submission of a single set of biometric samples to a system for identification or verification. Some biometric systems permit more than one attempt to identify or verify an individual" (Source: Certified Biometrics Learning System, Module 2, Biometrics System Design & Performance, Copyright 2010 IEEE, pp. 3–7).

It should be noted that there are two specific types of situations that can lead to a repeated number of biometric trait attempts by the end user, and they are as follows:

1. The biometric trait that is presented to the sensor is rejected because it is deemed to be totally unreadable (as reviewed previously in this chapter);
2. The biometric trait that is presented to the sensor is ultimately accepted and processed successfully into a verification template, but a non-match is still declared by the biometric system because there is no current enrollment template that exists for the user in the system's database.

The biometric system can keep taking samples of the physiological or behavioral trait literally, forever. If this occurs, the end user will probably just quit the entire process of verification and/or identification altogether. The end result will be that the end user will be non-cooperative for future verification and/or identification attempts which will be required of them in order to gain access to the resources they are requesting.

Therefore, it is very important for the system administrator to set the appropriate quality threshold of the biometric system so that a balance is struck in the number of attempts which are required by the end user. In other words, it is a fine line that needs to be maintained between end user perception and convenience versus the security of the business or organization.

But, this balance comes with a sharp tradeoff:

1. Requiring more attempts on the part of the end user will increase both the Failure to Accept (FAR) and the Failure to Enroll (FTE) rates;
2. Requiring a lesser number of attempts on the part of the end user will greatly increase the False Match Rate (FMR) and the False Non-Match Rate (FMNR).

Finally, there are three other factors which affect the sensing and data acquisition subcomponent:

1. Illumination: This is the dedicated light source that exists from within the biometric device. This light source flashes onto the physiological trait so that the raw image can be captured, but the amount and the intensity of light that is used varies greatly from technology to technology (a lot of this is due to compensate for the extreme changes in lighting in the external environment, in order to capture a consistent, quality raw image);
2. Contrast: This refers to the differentiation that exists between the lightest and the darkest portions of the biometric raw image which is ultimately captured. It can also refer to the total number of dark pixels versus the total number of light pixels in the raw image;
3. Background: All biometric systems keep the background image to a solid, light color.

## SIGNAL AND IMAGE PROCESSING

The next subcomponent to be examined after the sensing and data acquisition is the signal and image processing subcomponent. This is the part of the biometric system that consists of highly specialized mathematical algorithms that detect, enhance, and extract the unique features from the raw image that was captured in the previous step just described.

These very sophisticated mathematical algorithms take these unique features and convert them to the appropriate mathematical file type. This in turn becomes the enrollment and the verification templates. The signal and image processing subcomponent consists of the following steps:

1. Preprocessing of the biometric raw image;
2. Various types and kinds of quality control checks;
3. Biometric raw image enhancement;
4. Unique feature extraction from the raw image;
5. Postprocessing of the unique features;
6. Unique feature compression and decompression.

## Preprocessing of the biometric raw image

The first portion of the signal and image processing step is the actual preprocessing phase. In this phase, the biometric raw image is automatically fit into a standardized orientation, the center point is identified, any extraneous data is segmented out, and fit to be contoured into a certain set of established permutations.

Within the preprocessing phase, there are numerous subphases which occur, and they are as follows:

1. Detection: This refers to determining if any type or kind of a biometric signal exists in the data stream, which is captured from the biometric raw image. For all of the existing biometric modalities (both physical and behavioral-based), the detection of some sort of a signal will occur, because the physiological/behavioral trait is presented consistently and voluntarily to the sensor;
2. Alignment: This involves determining the proper orientation and centering of the biometric raw image. With the former, this includes taking into account any types and kinds of differences and variances in the specific rotation of the biometric raw image. This step is very important, as it can increase the accuracy of the unique features that are captured, as well as the quality of the subsequent enrollment and verification templates that are created. With the latter, the center of the biometric raw image is ultimately determined, and based on this, the image is then repositioned;
3. Segmentation: This part of the preprocessing phase can be specifically defined as "the process o parsing the biometric signal of interest from the acquired data system" (Source: Certified Biometrics Learning System, Module 2, Biometrics System Design & Performance, Copyright 2010 IEEE, pp. 3–11).

Before the segmentation subphase can occur, the detection subphase must be accomplished first successfully. After all, in order to actually segment the biometric raw image, there must be proof of actual data in it, which is what the detection subphase exactly accomplishes.

From the biometric raw image, once the needed data is located, the extraneous data, or the data that is not needed is then cropped into the segmentation subphase. This subphase can also be applied to video sequences as well, especially when facial recognition is being used in conjunction with CCTV technology.

For example, only the relevant scenes where the face distinctly appears can be included, with the other irrelevant scenes being cropped out appropriately. The next subphase after segmentation is what is known as the normalization subphase. This step involves adjusting or scaling of the relevant biometric data which is discovered in the last subphase. This is then contained within a range of predetermined values which are deemed to be acceptable.

Normalization can literally increase or decrease the levels of the signals which have been detected previously, and also enhance that particular signal so it can be processed more efficiently by the biometric system in subsequent steps. There are two different types of normalization:

1. Energy level normalization: This process subtracts the effects of differing levels of luminescence by increasing or decreasing the contrast amongst the features. Also, this process is usually applied to regions with the detected signal, and grayscale normalization decreases the differences found in the gray levels of the biometric raw image;
2. Scale/shape normalization: In this process, the signals are exactly modified so that easy comparisons can be made. Also, biometric raw images that have complex shapes or curvatures can also be adjusted with normalization.

## Quality control checks

The next subphase in the signal/image processing is quality control, and it consists of the following components:

1. Assessing the level of quality in a feedback loop;
2. Using quality control as a predictive mechanism for measuring how well biometric templates will compare with another (in particular, the enrollment template versus the verification template);
3. Assessing the level of quality control as an input mechanism from within multimodal biometric solutions.

In the first component (quality control in a feedback loop), it is very important that the biometric system detects any flaws in the signal which is captured in a very quick and efficient manner. In this particular instance, if a poor signal is determined, the feedback loop mechanism in the biometric system is automatically triggered, and the end user is asked by the system to present their particular biometric trait once again.

This particular mechanism is especially useful for the non-cooperative type of end user, whose biometric trait has to be presented and repeated numerous times.

The second component (quality control as a predictive mechanism) is very often used to gauge as to how well the two templates (enrollment and verification) will compare against one another. In these types of scenarios, very often, quality scores are calculated and utilized. The quality score serves four distinct purposes:

1. If the signal can provide enough data so that the two templates can be matched against one another;
2. The quality control score can also be utilized as an indicative tool to measure and ascertain just how much signal and image processing is really required by the biometric system;
3. The quality control score can be used to formulate a specific type and kind of matching strategy (for example, if both sets of irises will be scanned or just one iris will be scanned, or if two fingers will be scanned versus all ten fingers);
4. The quality control score can serve as a measurement technique as well to determine if the needed or required amount of unique features have actually been extracted in order to create quality enrollment and verification templates.

In the third component (quality control as an input mechanism), the quality control score can be used to "meld" and bring together biometric information and data from the various biometric modalities that exist in multimodal biometric systems.

In such solutions, quality control scores can also be utilized to create different weighted values in the signal and imaging subphase of the different biometric modalities which are involved in the overall security solution.

## Image enhancement

The next subphase after the quality control subphase in signal and image processing is that of the enhancement subphase. This involves taking the signal (which has been just previously discussed), and enhancing and/or further processing it for further use later by the mathematical algorithms which create the enrollment and the verification templates.

The enhancement subphase involves the following components:

1. The replacement of the unreadable features of the biometric raw image with blank spots, so that only clean and distinct unique features will be extracted;

2. Taking the very complex parts of the biometric raw image and simplifying them further for easier processing by the biometric system;
3. The fixing and repairing of any damaged or missing features from the biometric raw image. In other words, this is the extrapolation and calculation of what a feature could very possibly look like, based upon the careful examination and analysis of adjacent features (this is a must-the degraded feature(s) must be next to, or adjacent to a reliable and distinct feature).

## Feature extraction

After the enhancement subphase comes the feature extraction subphase in signal and image processing. This is also known as representation, and this is the very core of it. Feature extraction occurs at both the enrollment verification and/or identification stages.

It consists of the following methodology:

1. The unique feature location and determination;
2. The extraction of the unique features;
3. The development and creation of the enrollment templates;
4. Behavioral model creation;
5. The development and creation of the verification templates.

With unique feature location, this simply involves the examination of the biometric raw image of the distinct features that set them apart from the other features. For example, a fingerprint raw image would involve the location of the minutiae within the ridges and valleys, and an iris raw image, would involve the location of the furrows, coronas, freckles, etc. which exist from the iris itself.

Extraction simply involves, via the use of the appropriate algorithm, the removal of the unique features that will be utilized. With verification template creation, this simply involves taking the unique features that were removed previously and converting them over to the verification template, which is nothing but a mathematical representation of the unique features. For example, with the minutiae of the fingerprint, it would be represented by a binary mathematical file (which is a series of zeroes and ones).

In terms of behavioral model creation, statistical models are created to represent unique features. As its name implies, this is used primarily for the behavioral-based biometrics of signature recognition and keystroke recognition, where: (1) There are no physiological raw images to be captured; and (2) Behavioral-based biometrics are much more dynamic in nature versus physical based biometrics, where the raw images are much more static in nature.

Finally, with the creation of the enrollment templates, the same processes are used in the creation of the enrollment templates, but this time, the verification template is compared with the enrollment template to determine the closeness of the two templates. Subsequently, the result of this comparison is used to either grant or deny access to the individual in question who is requesting the use of a particular resource.

Specifically, it is the feature extraction algorithms that take all of the unique data that is extracted and convert them over to the appropriate mathematical files, which thus become the enrollment and verification templates. While the goal here is to maximize the time spent on determining and ascertaining the distinctiveness of a biometric raw image and less time on the extraction and matching processes, there is a distinct trade-off to be made.

For example, the less time is used in analyzing the distinctiveness of the biometric raw image, the more time is needed for unique feature extraction and matching of the enrollment and the verification templates. Finally, a good biometric raw image is one that has a lot of distinctiveness and uniqueness to it, and the physical structure of it does not change much over the lifetime of an individual.

## Postprocessing

The next subphase after feature extraction is postprocessing. This occurs when the unique features are extracted. The postprocessing subphase looks for any groupings of the unique features that may have occurred in the extraction process (any grouped unique features can cause errors and increase the processing time later) or for any poor quality features.

Also, during this subphase, the biometric raw image is either discarded or archived, from the biometric system. But there are issues to be considered. For instance, retaining any type or kind of biometric information and data can raise serious privacy rights issues, and if any of it is discarded by any means, it can cause a serious hindrance for future upgrades in terms of both biometric hardware and software for the future (for instance, if the biometric information and data are indeed discarded, the end users may very well have to re-enroll their templates yet once again if the newer biometric system is ever set into place).

## Data compression

The very last subphase in signal and image processing is known as data compression. This subphase is used primarily in two distinct scenarios:

1. If the biometric template or data takes up too much memory in the biometric system;
2. If the biometric template or data absolutely must have to travel across a low network bandwidth channel.

The basic premise of data compression is to bring down the biometric data template or data into a neat and manageable size while keeping the essential information and data that is needed for verification and/or identification intact. Data compression is deemed to be successful if the template or data, after it has been decompressed, can function in the exact same manner as it was intended to before it was compressed down.

There are two extremes associated with data compression, and they are known as "lossy" and "lossless". With the former, this creates an irreversible loss of all of the biometric template information and data, and the latter creates a situation where all of the biometric template information and data can be recovered after it has been compressed down.

It should be noted that most compression methods fall right in the middle of this spectrum. Also, a higher compression ratio will result in a much higher degradation of the biometric template information and data. This is so because some of the biometric information and data is either rounded up or down, and the original information and data are usually not retained. To counter this, neural network technology can be implemented to recreate any lost or damaged biometric template information and data after it has been compressed down.

## Data storage

After the signal and image processing subcomponent comes the data storage subsystem. This subsystem can be viewed as the database of the biometric system and is thus primarily responsible for the storage of all biometric information and data, including both the enrollment and verification templates. Specifically, a biometrics database can be defined as: "A collection of one or more computer files. For biometric systems, these files could consist of biometric sensor readings, templates, match results, related end-user information, etc." (Source: Certified Biometrics Learning System, Module 2, Biometrics System Design & Performance, Copyright 2010 IEEE, pp. 3–24).

A concept associated with a biometrics database is that of a gallery. A gallery can be defined as: "The biometric system's database, or set of known individuals, for a specific implementation or evaluation experiment" (Source: Certified Biometrics Learning System, Module 2, Biometrics System Design & Performance, Copyright 2010 IEEE, pp. 3–24). Simply put, a gallery is a database of images, such as that of what a facial recognition system would use (primarily for storing Eigenfaces, and CCTV camera footage).

Another major function of the biometrics database other than for storage is that of what is known as search and retrieval, which is a direct function of the matching process when the enrollment and verification templates are compared with another. For example, if identification (1:N) is the primary application scenario of the biometric system, then the entire database must be searched.

In the above type of scenario, a search database is then created from the master biometric database, in an effort to expedite the searching process. This secondary database (the search database) is created in order to protect the master database, by searching through the various partitions of it.

## Search and retrieval techniques

With a biometrics database, there are two types of search and retrieval which can occur, and they are as follows:

1. Verification: This is when the database is searched for a claimed identity. This is a very straightforward process because only one matching record has to be found, which in turn leads to less processing time imposed onto the biometric system;
2. Identification: This is when the entire database must be searched completely, because of the lack of a claimed identity. This can be a very complex process, time-consuming, and can severely constrain biometric system resources.

Search and retrieval techniques in biometrics-based databases are affected by the following factors. In other words, the following variables can greatly affect the processing of the biometrics database:

1. Winnowing errors: Although the advantage of winnowing a database (which involves the filtering, binning, and indexing of records within a database) is designed to optimize the search space in a database (by creating the temporary partitions in it), it can also lead to search and retrieval errors;
2. Differences in the quality of the biometric data in the biometrics database: Search and retrieval techniques are also greatly affected by the differences in the quality of data which exist in varying types and kinds of multiple databases (for instance, those found in multimodal biometric systems). In order to compare data efficiently and successfully in these heterogeneous environments, the biometrics database system must be able to quantify the performance for ensuring good-quality comparisons of the differing biometric templates. As a result, image processing algorithms are introduced to help fuse the biometric raw images together of the different modalities;
3. Multiple types and kinds of biometric data: In multimodal biometric systems, multiple biometric templates can be searched for simultaneously. Although this is designed to greatly decrease the

search and retrieval time required, it also puts an extra burden on the processing resources of the biometric systems in a multimodal environment;

4. Biometric template accuracy requirements: If further granular detail is required of the biometric templates, the biometric system can easily extrapolate such features via the mathematical algorithms it possesses. But again, this can also increase search and retrieval time within the database and be a further constraint on the processing resources.

## DATABASE SEARCH ALGORITHMS

To search a biometrics database effectively and efficiently, various types and kinds of search algorithms can be used, such as the following ones:

1. Sequential searching: This involves examining and analyzing biometric templates for the first through the last records in a chronological or sequential order;
2. Binary search trees: With this technique, a directory root tree structure is placed in a specific location within the biometrics database. This thus creates two smaller subtrees. The biometric templates can then be searched for in one subtree or the other subtree;
3. Hashing: This method uses a specific key which is assigned to each biometric template in the database, to allow for quick searching. This is especially useful in verification (1:1) applications;
4. Parallel searching: This is where multiple biometric templates are searched for at the same time in multiple biometric databases.

Another very important element of the biometrics database is the actual server upon which it will reside and be hosted. Typically, in small- to medium-sized applications, the actual biometrics database resides within the biometrics database itself. But, in much larger applications, especially that of client–server types, the biometrics database resides on a separate, physical server.

As a result, the choice of hardware thus exists in which to store the biometrics database, and they are as follows:

1. Virtual servers: These are typically stored at the physical premises of the Internet Service Provider and are often the most economical choice, due to economies of scale;
2. Blade servers: This type of server can help with hardware standardization and allow for the biometrics database applications to be scaled upwards quite easily and rapidly.

An image of a Blade Server is illustrated in Figure 3.2.

*Figure 3.2* This is an image of a blade server, where they are compact enough that they can be stacked on top of one another. (A close up of a metal object on a wooden post photo – free image on Unsplash.)

Of course, an equal, important component along with the choice of hardware is the software that is used in the design and creation of the biometrics database. A lot will depend upon the requirements of the end user, but most typically, either closed-source or open-source database software can be utilized in development.

With the former, the choices range from SQL Server to Oracle database software, and with the latter, the choices range from MySQL to PostGRESql database software. Also, along with the choice of the appropriate hardware and software, the following are just as equally important factors that need to be taken into consideration before the actual biometrics database is created and built:

1. The requirements of the biometrics systems based upon the system analysis and design;
2. The byte size, the total number, and the kinds and types of biometric templates that will be stored in the biometrics database;
3. The exact biometric application which will be used (such as either identification or verification);
4. The expected increase in size of the biometrics database (thus, scalability will be very important here);
5. The speed and efficiency of the search and retrieval queries for particular types and kinds of biometric information and data;
6. Any privacy rights issues which may arise with regard to the storage of the biometric templates in the biometrics database;
7. The overall network security of the biometrics database;
8. The overall stability of the biometrics database in the long term.

## Backup and recovery of the database

After the hardware and the software choices have been decided upon in order to build the biometrics database, the next factor that needs serious consideration as well is the backup and recovery of the biometrics database, in case of either physical or natural disasters. The considerations in this factor include:

1. The detection and repair of any types and kinds of errors in the biometrics database;
2. The type of schedule implemented for the backup of the biometrics database;
3. Any offsite storage locations for the biometrics database;
4. Any type or kind of redundancy (this primarily includes the choices of both fail-safe and fail-over options);
5. The backup and recovery plan in case the biometrics database is either lost or heavily damaged due to physical or natural disasters.

Finally, it should be noted that most biometric systems use a combination of storage mechanisms just described in order to ensure continued operations in case of either biometrics database failure or network connectivity failure.

## Database configurations

In terms of biometrics database configuration, there are four different types of setups that are available, and they are as follows:

1. A centralized database;
2. A decentralized database;
3. A client database;
4. A portable storage medium.

With a centralized database structure, the biometric information/data as well as the biometric templates are stored in one database, or a centralized database, as the name implies. This central database is then further networked to all of the biometric devices. As a result, once the biometric templates get processed at the device level, they get transmitted across the network media into the central database for storage.

Also, if biometric template transaction processing occurs at the central database server level, the server then becomes known as a "biometric recognition server". A key advantage to this type of structure is that remote identification and/or verification is possible.

In terms of a decentralized database structure, multiple database servers are required to store all of the biometric information/data as well as the biometric templates. However, the actual database servers can reside at different locations, many miles away from one another, or they can all be at the same location.

Alternatively, another type of configuration in this particular structure would be to have the central database divided into separate partitions, and then these partitions would become literally their own biometrics database, which can then be stored at different locations. A prime advantage of this type of configuration is that the network processing time is much quicker, and the overall advantage of a decentralized database structure is that of redundancy of the biometric information/data, and the biometric templates.

With regards to a client database, this refers to storing the actual biometrics database in the biometrics database itself. This type of configuration is very often used in stand-alone biometric devices, with the prime advantage being that of very quick processing times for either identification and/or verification types of applications.

However, if there are many biometric devices involved in the configuration, then these devices can be then be connected to a central server, which thus forms a client–server network topology. All biometric transaction processing and storage, and all of the biometric information/data and the biometric templates occur and are stored at the server level. However, the central server does keep the client biometric devices updated when verification and identification transactions do occur.

Finally, with regards to a portable storage scheme, this simply refers to the fact all of the biometric information/data and the biometric templates are stored on a device that is actually mobile in nature, and by design. A prime example of this is the smart card, which consists of a memory chip, upon which the biometrics database can be stored.

A key advantage of this type of storage scheme is that the end user is in charge of his or her own smart card, and all that is needed is a smart card reader to read and process the biometric templates that are on the smart card. Thus, the issues related to violations of privacy rights should be greatly mitigated, as well as the feeling of non-intrusiveness greatly increases as well with the use of a smart card.

## TEMPLATE MATCHING

After the data storage subsystem, the next subsystem, the next subsystem in the biometrical architectural design is that of the matching subsystem. This is also a very important subsystem, as this is the component that compares the enrollment templates against the verification templates, for either identification or verification types of scenarios.

Specifically, matching is defined as "the process comparing a biometric sample against a previously stored template and scoring the level of similarity. Systems then make decisions based upon on this score and its relationship (above or below) a predetermined threshold" (Source: Certified Biometrics Learning System, Module 2, Biometrics System Design & Performance, Copyright 2010 IEEE, pp. 3–32). It should be noted that the matching subsystem can search an entire database or winnowing techniques can also be utilized.

Rather than searching the entire database, winnowing techniques search only a particular subset or smaller partition of the actual database. To use winnowing techniques effectively, the concept of penetration rates becomes very important to understand. Specifically, the penetration rate can be defined as "a measure of the average number of pre selected as a fraction of the total number of templates" (Source: Certified Biometrics Learning System, Module 2, Biometrics System Design & Performance, Copyright 2010 IEEE, pp. 3–32).

Of course, the lower the penetration rate, the better, which simply means that less biometric information and data have to be searched through. To help reduce the penetration rate and to make winnowing as effective as possible in searches and retrievals, the concept of binning is also used. The primary goal of binning is to create and develop a much smaller search space, in order to help increase the accuracy rate.

With the techniques of binning, permanent categories on the biometric information and data are created, and these categories are as follows:

1. Exogenous data: This is data that is related to the biometric template but is not necessarily derived from it;
2. Endogenous data: This is data that is directly derived from the biometric template (such as the minutiae from the fingerprint).

Binning, with the use of pre-selection algorithms, can either filter and index through the smaller partitions of the biometric information and data. Specifically, a pre-selection algorithm can be defined as "an algorithm to reduce the number of templates that need to be matched in an identification search of the enrollment database" (Source: Certified Biometrics Learning System, Module 2, Biometrics System Design & Performance, Copyright 2010 IEEE, pp. 3–33).

With filtering, the search is conducted within the entire database, and both the endogenous and the exogenous biometric information and data are used to find the biometric templates whose data exactly matches the query that was executed. Filtering can also be performed in stages, where both the queries and the data become more refined and specific in the biometric information and data that is being sought for.

Indexing involves the assignment of a numerical value to each piece of biometric information and data in order to help decrease the search space and increase the retrieval speed. In the end, only those biometric templates whose indexes are close to one another are searched for, which subsequently increases the search and retrieval time.

It is the matching subsystem that does the comparing and evaluating between the enrollment and the verification templates. Although it appears quite easy and simple to the end user, the matching subsystem is quite a very complex process, because the laws of statistics and probability are used in order to determine an actual match.

There are three types of matching scenarios which can occur, and they are as follows:

1. Verification matching;
2. Identification matching;
3. Vector matching versus cohort matching.

With verification matching, as it was closely examined in Chapter 1, the goal is to confirm the claimed identity of a particular individual. In other words, the question of "Am I whom I claim to be?" is being answered by the biometric system. This is known as one-to-one matching, or mathematically put, it is a 1:1 relationship. In this scenario, only one enrollment template is compared against only one verification template.

With identification matching, as it was also reviewed in Chapter 1, the goal is to confirm the unclaimed identity of a particular individual. In other words, the question of "Whom am I?" is being answered by the biometric system. This is known as one-to-many matching, or mathematically expressed as a 1:Many relationship. In this scenario, the verification template is compared to many (perhaps even the entire database) enrollment templates in order to ascertain the true identity of the individual. Identification is also known as "biometric searching".

Also, matching can occur in both vector measurements and cohort matching. With the former, vector scores are computed and calculated and are subsequently further broken down into Euclidean (geometric) distances. If the enrollment and verification templates are deemed to be close enough in terms of this distance by the biometric system, then the individual is authorized to gain access to whatever resources he or she has requested.

With the latter, statistical models are actually created based on the enrollment templates which were constructed prior. The matching process then involves a determination by the biometric system on whether or not the particular signal could have been generated by that certain statistical model. The score is then calculated by using statistical probabilities, and if the score is close enough, then the individual's identity will be confirmed.

Finally, in a verification application, the result is often displayed by the system as a numerical score, and with identification applications, the result will be displayed also as a numerical score. Instead, the identification result is shown in a range from zero to some upward number, as it is determined by the system administrator. It is then up to the decision-making subsystem within the biometric system to confirm or deny the identification of the individual in question.

## Threshold decision-making

The next step after the matching component is the decision subsystem. Now, that there are enrollment and verification templates that can be matched, the next process is to examine the two templates, evaluate the closeness of similarity or the degree of dissimilarity between the two, and from there make an actual decision if the individual should be authorized to gain access to the resources that he or she has requested.

Specifically, in terms of biometrics, a decision can be defined as "the resultant action taken (either automated or manual) based on a comparison of a similarity score (or similar measure) and the system's threshold" (Source: Certified Biometrics Learning System, Module 2, Biometrics System Design & Performance, Copyright 2010 IEEE, pp. 3–36).

For example, if the comparison result between the enrollment and the verification template is above the threshold level, a match is considered to have happened. However, if the comparison result is below the threshold level, then a non-match is declared, in verification scenarios. It should be noted that the security threshold is normally established to maximize the statistical probability for a correct match to occur.

Also, at this point in the biometric system, the decision-making component is heavily dependent upon the previous subsystems as previously reviewed. The quality level must be high in these previous subsystems in order to render correct matches to occur.

In terms of identification scenarios, the entire database is scanned, and then a subset of the database is presented to the system administrator, of possible matched candidates. There are three techniques which a biometric system can utilize in order to return a subset of the biometrics database, which consists of the list of possible, matched candidates. The techniques are as follows:

1. Threshold-based: In this technique, every template in the biometrics database is scanned through entirely, and a comparison score is calculated for every candidate who is a potential match. A potential match is determined by whether this particular comparison score is above the security threshold level. However, if there are no comparison scores that exceed the security threshold level, then the identity of the person cannot be confirmed, because it is assumed then at this point they have never been enrolled in the biometric system;
2. Rank-based: With this technique, each and every biometric template is also scanned in the biometrics database, and a subset of possible matches is returned to the system administrator, with ranks ranging from best to worst;
3. Hybrid method: This is a combination of the threshold and ranking methods, and in this scenario, the most common application is to compile and return a ranked list of possible candidates whose comparison scores are well above the established security threshold level.

The security threshold upon which the score is compared against is not just solely decided by the system administrator, but rather, by a combination of factors. First, there are three different levels for the security threshold, for both identification and verification scenarios. They are as follows:

1. Static threshold: This occurs when the same security threshold level is established for all verification and identification transactions;
2. Dynamic threshold: This happens when the security threshold is changed because of variances in both the internal and external environment;
3. Individual threshold: These are static in nature as well, but can vary in the level of setting for different employee job roles (for example,

the C-Level Executive of an organization will have a different security threshold level than that of a temporary worker).

Second, the variables used in deciding upon the security threshold setting include the following:

1. The quality score (the percentage of a biometric raw image that is actually deemed to be usable by the biometric system);
2. The summation of other compiled results from different modalities within a multimodal security solution;
3. Any additional information and data which exist upon the comparison score.

Third, there are two different subset elements of the above variables which help to further define what the exact security threshold level should be, at a much more granular level. These two elements are:

1. Lights on mode versus lights out mode;
2. The system policy which is set forth for the particular biometric technology.

With the first subset element, lights on mode refer to the fact that some type or kind of human intervention is required, along with the decision made by the biometric system in order to confirm or deny the identity of the individual in question. With the lights out mode, the biometric system automatically makes the decision itself, without any human intervention.

With regards to the system policy sub-element, this decides whether the match result should be ultimately accepted or rejected. With regards to identification scenarios for this particular sub-element, this also helps to decide whether a matched candidate should be accepted or rejected, due to the sheer volume of potential candidates which can be returned back. Finally, it is the system policy sub-element that helps to decide as well the characteristics of a biometric template which is desired in order to fit the system requirements.

## Administration decision-making

After the matching subsystem, the next subsystem to be encountered in any type or kind of biometric system is that of the administrative system. It is at this juncture in the biometric device in which human intervention is possible, decision-making by a human can also happen, and database monitoring can transpire as well.

The administrative subsystem can be as simple as a keypad on the biometric device itself, or it can be as complex as an entirely separate software application in a biometrics-based, client–server network topology. Despite what the actual interface looks like, there are a number of critical functions in the administrative subsystem serves which are as follows:

1. The communications role in the biometric system;
2. Biometric system control and administration;
3. Biometric system auditing, reporting, and control;
4. Keeping track of the biometric system with different mode controls;
5. Biometric system administrative controls.

It should also be noted that the administrative subsystem is designed to act as the primary interface to that of the overall security system in general, as well as the actual hardware, as mentioned previously. For example:

1. Biometric devices are rarely just stand-alone devices anymore. For instance, they very often networked together with other types and kinds of biometric modalities, thus serving a much larger security application. This can include physical access entry, logical access entry, time and attendance, and perhaps even a combination of all three;
2. But, if a biometric device does indeed have to operate as a standalone device, all communications with it are still conducted through it at the administrative subsystem.

But apart from the internal security communications just described up above, the administrative subsystem is used also to allow for the biometric system to communicate with the other types and kinds of security devices which are non-biometric related in nature, such as CCTV cameras, background checks technology, and door alarms.

It is critical at this point that the administrative subsystem be configured to exactly meet the needs of the system requirements of the overall biometric architecture. Some of these configurations include the following:

1. The formatting of the biometric templates (both enrollment and verification), so that they are in compliance with the various standards as also described earlier in this book;
2. The compilation into a report format of all biometric system statistics and logs to the system administrator. Such data can items like the various error rates that have been encountered over a certain time period, the differing types of biometric transactions that have occurred, authentication/non-authentication of end users, etc.;
3. The reformatting and transmission of biometric templates which are to be sent to other biometric devices, such as those in a multimodal security solution.

The administrative subsystem provides the system administrator the authority to control all aspects of the biometric system, whether it is a standalone device or a multimodal biometric security system. However, there are a number of key controls which are deemed to be the most critical, and they are as follows:

1. Biometric templates adaptation;
2. The establishment of the security threshold values;
3. Reporting and control;
4. System mode adjustment;
5. Granting privileges to end users.

## Biometric templates adaptation

With the biometric template adaptation control, this merely refers to the fact of how often the end user needs to reenroll their biometric templates, or update them, due to the fact that the templates can degrade over time. It should be noted that it is not the biometric template in the database that erodes over time, rather, template aging refers to the fact that the end user's physiological and biological features can degrade over time, thus causing a mismatch between the newly created verification template which has been just presented, versus the older enrollment template, which has been captured and created at an earlier point in time.

Specifically, template aging can be defined as "as the increasing dissimilarity between a biometric reference and the actual biometric trait of the user due to periodic changes" (Source: Certified Biometrics Learning System, Module 2, Biometrics System Design & Performance, Copyright 2010 IEEE, pp. 3–43).

There are two methods to alleviate the problem of template aging, and they are as follows:

1. Enrollment: If there is a huge difference between the enrollment template and the newly created verification template, all that is required is for the end user to present their particular biometric trait so that a new enrollment template can be created;
2. Template adaptation: If there is not too much of a difference between the two templates, then the enrollment template is merely "dynamically update" (or reconstructed) so that the templates (the enrollment and verification) are more closely aligned with one another.

## ESTABLISHMENT OF THE SECURITY THRESHOLD VALUES

With regards to security threshold establishment control, this refers to the security threshold values at which the end user will either be ultimately accepted or rejected by the biometric system. The exact threshold can be established by setting at a static, dynamic, or individual threshold level, as examined previously in this chapter. This is probably one of the most difficult decisions to make after the actual biometric system has been installed, and the system administrator needs to take this into great care.

## Reporting and control

With regards to the reporting and control mechanism, this functionality provides the system administrator the ability to configure and prepare reports with regard to the biometric system in general. Some of the options that the system administrator has at their disposal include:

1. The specific data which needs to be recorded;
2. How long the data need to be archived;
3. How often (the exact frequency) that the reports need to be published;
4. Configuring the reporting tool(s) so that the relevant and appropriate data is presented to the system administrator.

## System mode adjustment

In terms of the system mode adjustment control, there are three separate and unique modes which the system administrator can set:

1. Training mode: This occurs when the biometric system is used to teach the end user about biometrics, how the biometric system works, and how to enroll their respective templates properly;
2. Testing mode: This happens when technological (including both hardware and software) upgrades are performed on the biometric system;
3. Operating mode: This happens when the biometric system is functioning normally, as required.

## Privileges to end users

Finally, in terms of the privilege control, with the administrative subsystem, there are two types of settings that can be established, and they are as follows:

1. Access control and security settings;
2. Biometric information and data privacy settings.

## Data transmission

The final subsystem in the biometric architecture model is that of the data transmission subset component. With this subsystem, all of the other subsystems just discussed are networked together, and thus, bring the biometric system into one, cohesive working technology. However, it should be noted that this final subsystem is a rather broad one, and the networking involved includes all hard-wired and wireless networking.

The data which is transmitted in this subsystem includes primarily the following:

1. The biometric raw images;
2. The enrollment and the verification templates;
3. Other types and kinds of data include the authorization/non-authorization decisions made by the biometric system with regard to the end user.

Just like the traditional OSI model which will be reviewed later in this chapter, the data transmission subsystem performs the encoding of the biometric templates and other biometric information/data as it is outgoing from the biometric system, and also performs the decoding of the biometric templates and other biometric information/data as it enters into the biometric system.

One of the biggest weaknesses of a biometric system lies within the transmission of information/data that transpires. One of the biggest threats posed to the data transmission subsystem is that of channel effects. Specifically, it can be defined as "the changes imposed on the presented signal in the transduction and transmission process due to the sampling, noise and frequency response characteristics of the sensor and transmission channel" (Source: Certified Biometrics Learning System, Module 2, Biometrics System Design & Performance, Copyright 2010 IEEE, pp. 3–46).

At the present time, such tools as encryption and digital signatures can be used to combat this threat. Another emerging tool is that of Biocryptography, which will be reviewed completely in a later chapter in this book, Chapter 4.

The last section of this chapter reviewed extensively the system architecture that is involved in a biometric system. This architecture is technology neutral, meaning that, the system architecture model is applicable to all types and kinds of biometric technologies, whether it is physical or behavioral-based.

## BIOMETRICS PROJECT MANAGEMENT

Each and every biometric device, whether it is a small single sign-on solution that fits onto a netbook or something as large as a hand geometry recognition device, needs all of these subsystems in order to function, and to perform the necessary verification and/or identification transactions.

But this biometric system architecture model is often viewed as more of a theoretical one to the C-Level Executive, especially to the CIO and the

CFO. What matters to these top-level executives is the actual project implementation of the biometric system after it has been procured and financed. In other words, before the biometric system can operate in full mode, some key questions need to be answered, such as:

1. How will the biometric system be installed at the place of business?
2. What kind of after-implementation testing will take place in order to ensure that all of the security requirements have been met by the biometric system?
3. More importantly, how will the end users be trained in order to effectively use the biometric system?

All of these questions and much more obviously need to be answered before the biometric system can into a live and fully functional mode at the place of business or organization. These questions can only be answered by examining the scope of the biometric system through project management techniques. As a result, this section of this chapter is intended to provide the CIO, the CFO, and any other C-Level Executive involved in a biometric system procurement decision with the basic project management tools they will need in order to ensure a successful and effective, long-term deployment.

Specifically, this section focuses on the following project management techniques:

1. Biometric system concepts and classification schemes;
2. Biometric system design and interoperability factors and considerations;
3. Conducting a biometric systems requirements analysis;
4. Biometric system specifications;
5. The various biometric system architectural and processing designs;
6. Biometric subsystem analysis and design;
7. Biometric subsystem implementation and testing;
8. Biometric system deployment and integration;
9. Biometric system maintenance.

## System concepts and classification schemes

Long before even the biometric modality and the respective hardware and software are decided upon, most C-Level Executives must first have a clear picture of what they truly want out of a biometric system at their place of business or organization. True, a C-Level Executive would like to have a biometric system that will provide an enhanced or even a brand-new layer of security, but it is also important to consider the other aspects that a biometric system can offer.

For instance, apart from providing physical and/or logical access, will time and attendance for employees also be a desired application as well? These questions and much more must be carefully considered and answered by launching four project management efforts, which include:

1. Conducting an exhaustive system design process;
2. Analyzing the system requirements with the appropriate development management lifecycle;
3. Defining the specific goals of the biometric system (also referred to as the system concepts);
4. Deciding for what application(s) the biometric system, specifically if it will be used for physical access entry, logical access entry, time and attendance, or perhaps even a combination of all these;
5. Also deciding what specific mode the biometric system will operate in, for example, verification or identification, or even both?

It is important that all of these questions be answered, and that nothing is left to mere and haphazard guesswork or even blind assumptions. The framework for this can be answered by using what is known as the "waterfall model", which is a very popular management methodology to utilize in technical and security-based implementations.

But, in order to first the systems concept and classification scheme component as previously mentioned, four individual processes must be accomplished, which are:

1. Acquire;
2. Supply;
3. Develop;
4. Operations and maintenance.

With the acquisition process, this is when the C-Level Executive performs the various RFI's (Requests for Information), the RFP (Requests for Proposals), and prepares the other types and kinds of procurement documents are necessary to help decide which biometric technology will be the best for their place of business or organization.

In terms of the supply process, this is when the biometric vendor(s) comes in. For example, the biometric vendor(s) can present their technology to the C-Level Executive in the previous process.

With regards to the development process, this is where the place of business or organization conducts the actual biometric system analysis and design, the system specifications, the biometric architectural design, and the testing of the actual biometric system before it goes live.

Finally, with the operations and the maintenance process, this is at the point where the biometric system is used for its intended purposes and applications, and must also be maintained as well, which includes monitoring the exact performance of the biometric system, and if deemed to be necessary, even migrating to a newer biometric technology and system, thus retiring the existing biometric system.

## Upgrading to a newer biometric system?

It should be remembered by the C-Level Executive that upgrading to a newer biometric system can be a very complex and expensive process and undertaking. One of the biggest issues for the C-Level Executive at this point is whether or not to retain the original biometric templates, as well as their related information and data.

The issues faced lie within the fear of the potential abuses of the biometric templates, at least in the eyes of the end users. But as always, the broader scope and the picture must always be kept in mind, and therefore, retaining the original biometric templates and their corresponding information and data, will make the transfer to a newer biometric technology and system just that much easier.

At the systems concept phase, a very high initial biometric systems design is created based upon primarily the needs, objectives, and the overall security mission of the place of business or organization. It is very important for the C-Level Executive to keep in mind that a biometric system is designed to serve a much bigger and broader security purpose.

The systems concepts phase includes the following components:

1. Security problem/needs identification and based upon that, the establishment of the high-level objectives;
2. Conducting and analyzing a feasibility study;
3. Creating and developing a high-level case study of the intended uses of the biometric system;
4. Defining the conceptual architecture of the proposed biometric system.

With regards to the security need/problem identification, in order to gain a clearer picture of the specific security needs of the place of business or organization, the following questions need to be asked:

1. What is the exact security need which needs to be identified?
2. Who will the end users be, and also, what are their exact, specific needs?
3. Who will be the system administrators of the proposed biometric system?
4. What are the restraining variables and factors that must be considered before the biometric system is deployed?

5. What level of performance is required by the proposed biometric system?

Although the chief reason for the procurement of a biometric system is for identity management, the needs can vary widely greatly under this broad umbrella of security. Once the above questions are actually answered, the goals for the proposed biometric system can thus be defined. The C-Level Executive must also remember that the goals for the proposed biometric system are strictly high level and have to correlate strongly with the objectives of the place of business or organization, as defined previously.

## The feasibility study

In terms of the feasibility study, this involves primarily conducting a cost analysis of the proposed biometric system. The following steps are included with it:

1. Getting a broad understanding of the existing security system at the place of business or organization;
2. The considerations of the of the alternative security solutions and costs in lieu of the proposed biometric system;
3. To what degree the existing security infrastructure can be leveraged, and how the proposed biometric system can fit into it;
4. The identification of the financial risk areas;
5. The development of the proposed budget and the financial constraints that go with it.

The C-Level Executive should also keep in mind that after the feasibility study is conducted, the benefits must be greater than the costs associated with the proposed biometric system in order for the place of business or organization to realize a positive Return on Investment (ROI).

## Application classifiers

Application classifiers are used in order to help further ascertain and drill down the biometric system and design. Classifiers are used in project management extensively today and are used to help ensure a successful deployment, whatever it may be. In terms of biometrics technology, the most commonly used classifiers are as follows:

1. Cooperative users versus uncooperative users: A cooperative user is one who is actively wanting to be confirmed by the biometric system, in order to gain access to the confidential resources and files that they

are seeking. This even includes impostors who are trying to break through the defenses of the biometric system. An uncooperative user is one who does not wish to be either identified or identified by the biometric system;

2. An overt biometric system versus a covert biometric system: An overt biometric system simply means that the end user is fully aware and cognizant that their physiological and biological traits are going to be captured by the biometric system, and that full cooperation is required on their part. A covert biometric system is where the end user is not at all aware of the biometric system, and that their physiological or biological traits are being collected by the biometric system. As a result, the end user is neither cooperative nor non-cooperative in this certain scenario;
3. Habituated users versus non-habituated users: A habituated user group is one which is expected to use the biometric system on a regular basis, without too much resistance or opposition. This also means that this type of user group can undergo a complex and in-depth training program if need be, about the biometric system they are using. With a non-habituated user group, this means that the end users are expected to use the biometric system very infrequently, and with much opposition. As a result, the biometric system must be very easy to use, in order to accommodate the very low learning curve of this particular end user group;
4. An attended biometric system versus an unattended biometric system: With an attended biometric system, human intervention is always required on a daily basis, especially to accommodate the uncooperative end user group. The enrollment and verification processes must always be supervised. With an unattended biometric system, no human intervention is required, whatsoever;
5. A constrained biometric system versus an unconstrained biometric system: With a constrained biometric system, the external environmental conditions remain identical on a daily basis for enrollment and verification of the end users. These conditions include the illumination, the type and kind of sensor used in the biometric system, and conditions in both the external and the internal environments. With an unconstrained biometric system, there are variations or differences daily in the environment in which both enrollment and verification occur in the biometric system;
6. Private users versus public users: Private users are those who have a vested use and are required to use the biometric system, such as employees of a place of business or organization. Public users are those that are not required to use a biometric system, and if they use it, it is very infrequent;

7. An isolated biometric system versus an integrated biometric system: With an isolated biometric system, the biometric templates and the corresponding information and data are never shared with other security systems. With an open, or integrated biometric system, the biometric templates and the corresponding information and data are shared with the other security systems, whether it is biometric-based or not;
8. A scalable biometric system versus a non-scalable biometric system: A scalable biometric system is one that can be upgraded or even replaced with newer technology, hardware, and software, in order to meet the increased demand. A non-scalable biometric system is one which works best for a fixed number of an end user population and cannot be upgraded or replaced if demand increases over time.

With regards to the above biometric application classifiers, a verification mode is much less taxing on the resources of a biometric system than that of an identification mode. This is so primarily for two reasons:

1. A verification mode only depends upon one single match result. An identification mode relies upon many matches, and the more matches that have to be made in order to confirm the identity of the individual, the greater the chances for errors and mistakes to happen;
2. A verification mode can be accomplished in just less than 2 seconds or so because only one match needs to be made. But with an identification mode, since many more matches have to be made, confirming the identity of an individual can take several minutes or more, thus rendering the biometric system infeasible in extremely quick processing time is an absolute, crucial requirement.

## System design and interoperability factors and considerations

Following the biometric systems and application classifiers, the next project management step is that of the biometrics system design and interoperability factors and considerations. This is part of the project management process where choosing how to develop the biometric system becomes a key question.

## COTS-based biometric systems

For example, a biometric device can be created within the actual place of business or organization (assuming of course, that such expertise does actually exist), or the development and creation of the biometric device can be outsourced to a third party, or the biometric device can be bought off the shelf, which is also known as "COTS". This stands for "Commercial Off The Shelf" biometric system.

However, whenever it comes to the decision of either making or buying the biometric device, cost and the least path of resistance to creation and development become the two driving factors. As a result of this, typically, COTS-based biometric systems are picked, because of the much lower purchase price which is involved, and the fact that a COTS-based biometric system is literally plug-and-play – just buy it, and install it. It should be noted that COTS-based biometric systems are also very technically advanced, and can be ready to serve market applications of physical access entry, logical access entry, and time and attendance, quite literally, out of the box.

Also, apart from the apart from the actual hardware, there is also COTS-based biometric software available, which is also plug-and-play. The benefit of this is that the cost is much lower than hiring a software development firm to design the biometric software application. When a COTS-based biometric system is acquired, it is often not interoperable with other types and kinds of biometric systems.

Interoperability is a must if the COTS-based biometric system is to interact with the other security systems at the place of business or organization, or if the C-Level Executive decides to create a multimodal biometric solution. There are two types of interoperability which are available for biometric systems:

1. The ability for other security systems or other biometric devices to be able to work with the COTS-based biometric system in full harmony;
2. The ability for all of the biometric devices and non-biometric devices to exchange biometric information and data readily.

The C-Level Executive should keep in mind that if the COTS-based biometric system is not interoperable, a phenomenon known as "vendor lock-in" can occur. This happens when a COTS-based biometric system cannot just be simply upgraded to a newer system, or rather, the entire biometric system must be replaced in order to modernize it with the latest technology.

If this does indeed happen, the costs and expenses can rise exponentially. But, if the COTS-based is designed and made to be interoperable, then simply upgrading is a much less expensive burden for the C-Level Executive. To help with the issue of interoperability, standards can be developed to help biometric systems become more integrated with one another, interchanged with another, and even upgraded over time.

## Proprietary or open-ended system?

However, standards can only go so far in helping with biometric system interoperability. Another key piece of the puzzle is the choice of whether to deploy a proprietary biometric system, or whether to deploy an open-ended biometric system. With a proprietary biometric system, the only choice that a C-Level Executive has is that if they want to upgrade their current biometric system, the existing one will have to be literally ripped out, and replaced with an entirely brand-new system.

Of course, this means great expense will be incurred by the place of business or organization. But of course, with an open-ended biometric system, this will possess a much higher level of interoperability than a closed system.

Thus, if a C-Level Executive wants to pursue an upgrade, all that will be required is a replacement of the hardware and software, which only be a fraction of the cost of installing an entirely brand-new biometric system. Also, upgrading a biometric system is much easier if open-source software is utilized and follows open-source biometric standards, such as BioAPI.

## Infrastructure assessment

Another important aspect of a biometrics system and interoperability factors and considerations is that of conducting an assessment of the infrastructure in which the new biometric system will be deployed. The infrastructure assessment involves a careful analysis of the following:

1. The physical infrastructure;
2. The platform infrastructure;
3. The network and directory infrastructure;
4. The database infrastructure.

With the physical infrastructure, this aspect must be considered very early on. This part examines the physical premises, power capacity, blueprint layout, as well as any types and kinds of enhancements that need to be made to accommodate the specific biometric system that will be deployed.

For example, special lighting may be required for a facial recognition system, or there could be different mounting heights that need to be considered for a fingerprint recognition device.

In terms of the platform infrastructure, this is also a crucial support that needs to be considered very seriously. For example, the chosen platform needs to be able to provide enough processing power (and of course, the more the better) in order to meet the high demands and speed of the biometric system, especially when it comes to verification and identification transactions.

The chosen platform must also be compatible with the hardware and the software that is being supplied by the biometrics vendor.

With regards to the network and directory infrastructure, it is also very crucial to have superb bandwidth, which will provide enough capacity and throughput for the quick transmission of the biometric templates. The C-Level Executive must remember that the point from where the biometric raw image is captured will be different from the point at which the verification and identification transactions and decision-making occur (this will most likely occur in a separate server).

Thus, there can be no bottlenecks whatsoever, in the network, if verification transactions are supposed to occur in less than 2 seconds. Also, if the biometric system is integrated or coupled into a directory structure (such as Microsoft Active Directory), the choice of communications protocols to be used also becomes important.

Finally, with the database infrastructure, this is where the biometric templates will be stored. So obviously, the choice of software and the design of the biometric system is of utmost importance. Also, key questions such as who will own and manage the database, and especially how the biometric templates will be protected, must be answered before the biometric system is procured.

## The human equation

Also, in this phase of the biometrics project management, the views of the people who are involved and also impacted by the biometric system need to be taken into serious consideration as well. In the end, it is not just the C-Level Executive who will be affected by the deployment of a new biometric system, but rather, by a whole range of people.

This group of people is known as the stakeholders and includes the following types of individuals:

1. The individuals who are responsible for the procurement and the acquisition of the biometric system (for example, the C-Level Executives, and other members of middle management);
2. The individuals who will monitor and run the biometric system (for example, the end users, and the IT support team);
3. The individuals who will be using the biometric system (for example the employees of the place of business or organization and other visitors).

It is important that during each phase of the biometric project, all of the stakeholders are involved in the entire process, and not to exclude any one group of individuals. That way, by being involved in the entire process, the acceptance level of the stakeholders for the biometric system will be much higher in the end.

Given the very important role of the stakeholders in the deployment of the biometric system, we now turn our attention to the more psychological aspects which need to be taken into consideration as well. Biometrics is one of those security technologies, when placed upon a spectrum with the other technologies, it has the most social impact upon end users and the public at large, as well.

Therefore, it is of crucial importance for the C-Level Executive to take into consideration all of the psychological aspects of the biometric systems, and the impacts it will potentially have upon all of the stakeholders. Although all of the psychological issues may not be resolved, if they are at least addressed from the outset, this will help to bolster end user acceptance in the end.

Some of the issues, both psychological and physical, that end users experience with biometric systems include the following:

1. Very different viewpoints and differences as to the exact human reactions to the concept of biometrics in general;
2. The varying levels of acceptance of using biometrics technology;
3. Any types or kinds of medical conditions or illnesses that could affect and impact both the physiological and the behavioral traits of the end users;
4. Manual labor jobs, which can affect the quality of both the physiological and behavioral traits (such as construction workers having their fingers impacted because of their job requirements).

## Ergonomic issues

A key component of the psychological issues of biometric systems is that of ergonomics. As it specifically relates to biometrics, ergonomics takes into consideration how to make a biometric system more comfortable for all of the end users to enroll their physiological and behavioral templates.

Some of these considerations include the following:

1. The height variances and differences among the end users;
2. The frequency of usage of the biometric system;
3. The design of the physical space of the biometric system by giving the end users a sense of privacy.

Another key aspect in addressing the psychological aspects of a biometric system is to foster open and honest lines of communication with the end users, especially the employees of a place of business or organization.

Allowing such a line of communication to be established will only help foster trust and acceptance of the biometric system by the end user.

## The feedback system

In this regard, feedback to the end user becomes very important. This can also be referred to as a feedback system. It can be represented by the following mechanisms:

1. Audio feedback in different languages in order to accommodate end users of different cultures and different countries;
2. Process pictures depicting to the end user how to easily enroll their physiological and behavioral traits into the biometric system;
3. Any type or kind of on-screen graphical user interfaces (GUIs) which are easy to comprehend and understand;
4. Other visual cues (also known as "passive feedback") to help guide the end users through the enrollment and verification processes.

It is best to accommodate the end users by offering a combination of the above feedback mechanisms. In this regard, as a result, all types and kinds of end users can thus, ultimately be served. Also, deploying a biometric technology can bring up fears and anxiety to the end users, because it is such a huge change.

This is where the appropriate change management mechanisms can as well, play an important role. Specifically, change management can be defined as "the process of preparing the organization for a new system" (Source: Certified Biometrics Learning System, Module 2, Biometrics System Design & Performance, Copyright 2010 IEEE, pp. 3–64).

To make change management easy and effective, as well as to help ensure a smooth transition, clear and open communications from upper management to the end users become extremely vital.

It is thus important that such communications start at a very early phase, perhaps even when the thought process of the consideration of a biometric system for the place of business or organization comes to the minds of the C-Level Executives. In this aspect, communications can take place in the following forms:

1. The advertising of the benefits of the new biometric system to the end users;
2. Addressing the privacy rights and civil liberties fears and issues the end users may have about the new and upcoming biometric system;
3. Keeping the end users informed that enrollment into the biometric system is not mandatory, other security options are available;

4. The pre-announcement of formal training sessions into the proper and effective use of the biometric system.

Because the biometric system has such strong impacts and influences on the end user, its acceptance and the ability to use a biometric system can actually have a much more pronounced, positive effect on system performance. On the converse, if the biometric system is poorly received by the end user of the place business or organization, then the system performance could quickly deteriorate in just a short matter of time.

However, if the place of business or organization is very large, say for example, with thousands of employees located in offices all over the world, then a study called "population dynamics" can be utilized to study the entire end user base, in order to create a more effective use of the biometric system.

## Population dynamics

Population dynamics is actually a statistical analysis that examines the average end user. Within population dynamics, two types of variability can be demonstrated:

1. Inter-class variability: This is the amount of uniqueness or the degree of distinctiveness amongst each end user. A much lower inter-class variability means that there could be potentially escalated rates of false acceptances (meaning an impostor is accepted and authorized by the biometric system), or even higher rates of FTE (this happens when the end user simply cannot enroll in the biometric system);
2. Intra-class variability: This is the amount of uniqueness or distinctiveness between repeat measurements of the same end user. If this particular variability is high, there could be an escalated rate of false non-matches occurring, during the verification or identification process.

Businesses or organizations that are very large (for example, over 5,000 employees who are located globally) will obviously have a large-scale biometric system implemented, as well as large-scale, biometric software applications.

Many C-Level Executives at this point question what is the exact criterion for what defines a large-scale biometric implementation at this phase in the biometric project management lifecycle. Although there is no hard and fast rule, most large-scale biometric systems and applications start at the range of 100,000 templates and greater.

However, not all biometric technologies are well suited for large such applications. There are two key variables that can pose a huge challenge when businesses and organizations try to implement large-scale biometric implementations. They are:

1. Accuracy;
2. Speed.

For instance, there is a negative correlation that exists between the above two, with regard to the scalability of a biometric system. As biometric systems and applications get larger in size, in terms of the total number of biometric templates that are stored and processed, the accuracy and speed decline at the same, if not greater, rate.

So the question now becomes of which of the biometric modalities are the best suited for large-scale applications. Here are some key points for the C-Level Executive to consider as they evaluate the upward expansion of their biometric system(s):

1. Utilize a biometric system which typically has a low FMR, these tend to scale up naturally to much larger biometric databases;
2. Implement a multimodal biometric system to help assure large-scale biometric database scalability;
3. Utilize the techniques of winnowing, such as filtering, binning, and indexing, to help with upward biometric system scalability;
4. Moving the existing biometric templates to the newer, much larger biometrics database, rather than having the end users re-enroll in physiological or behavioral traits. The end result of this is that this will save a lot of expense and time when scaling upwards;
5. Figuring out exactly the network throughput that is required when the biometrics database is scaled upwards;
6. Calculating the difficulty in scaling upwards the existing biometric system to the hypothetical largest biometric system that conceivably the place of business or organization could potentially support;
7. Trying to estimate the per unit cost of the biometrics database processing versus the total cost of processing the transactions which are needed for the entire end user population of the place of business or organization that is using the biometric system.

## Systems requirements analysis

The next step in the biometrics project management lifecycle is that of the biometric systems requirements analysis. Determining the requirements of what is demanded by a potential security system is one of the most important aspects of the biometrics project management lifecycle. For example, defining the specific requirements can be used to further pinpoint the security problems the place of business or organization is experiencing, and the goals that can be used to help allay those specific problems and risks.

The term requirement is a broad one, but for the purposes of biometric system deployment, it can be specifically defined as follows: "a condition or capability needed by a user to solve a problem or achieve an objective" (Source: Certified Biometrics Learning System, Module 2, Biometrics System Design & Performance, Copyright 2010 IEEE, pp. 3–69).

In biometrics project management, defining a set of requirements aids in the process of:

1. Clearly identifying absolute needs from unclear wants;
2. Ascertaining the specific functions and specifications of the new, potential biometric system;
3. Determining what the GUI will be, and how it will appear in terms of looks to the biometrics system administrator;
4. Providing a review of how the biometrics database will be designed, how it will function, and how it will integrate into the biometric system hardware and software;
5. Establishing the guidelines and the specific criterion for biometric system testing after it has been deployed and before it is made available to the end user in the place of business or organization;
6. Creating the biometric system documentation for maintenance, and more importantly, for end user training and use.

The biometrics system requirements consist of the following steps, and it will be reviewed in detail:

1. Biometric system requirements elicitation;
2. Biometric system requirement analysis and regulation;
3. Biometric system requirements documentation;
4. Biometric system requirements validation;
5. Biometrics requirements management.

## System requirements elicitation

With regards to the biometric system requirements elicitation, this is where all of the stakeholders who are involved in the biometrics project list out their specific needs and concerns about the potential biometric implementation. As a result of this process, end user acceptance of the biometric system should dramatically increase once the system is deployed, because the stakeholders will feel that they are taking an important part in the entire process.

Also, this process helps the technical staff (especially the IT staff) to pick up a biometric system that is specific to the security needs of the place of business or organization. It should be noted at this point that the biometrics requirements elicitation can either be passively or actively based. With the former, focus group interviews can be conducted with all of the stakeholders who are involved, and with the latter, process diagrams and other types and kinds of technical documentation are created.

Another effective tool which is utilized in this process is that of use cases. These are actually process flow diagrams that are used to map the biometric system in its entirety at the place of business or organization. To aid in this particular diagramming process, very often, the Unified Modeling Language, or UML, is widely utilized.

## System requirement analysis and regulation

In terms of the second phase, which is that of the biometric system requirements and negotiation, several steps are actually involved, which are as follows:

1. Rank requirements;
2. Model requirements;
3. Ascertaining if the potential, new biometric system can actually solve the security needs of the business or organization.

With rank requirements, the IT staff and the system administrators try to ascertain the statistical probability of severity if a particular system requirement were not to be implemented, for any type or kind of reason.

If the absence of a requirement causes a severe problem for the business or organization, then it is ranked with a very high statistical probability. However, if the absence of a requirement causes a less severe problem for the business or organization, then it is ranked much lower statistical probability. Also, cost functions associated with the biometric system can also be ranked on a more severe or less severe statistical probability rating as well.

With regards to the model requirements, as described previously, modeling languages are utilized to map out the potential, new biometric system. To assist with this task, two types of modeling languages can be utilized, and they are:

1. UML: As described previously, this is a visual design software package that is used to graphically design the various components of the entire biometric system;
2. Biometric and Token Technology Application Modeling Language (BANTAM): This is another visual design package, in which the hardware and software aspects of the biometric system can be mapped out quite easily.

Finally, in terms of ascertaining, if the biometric system can meet or fulfill the security requirements, this type of assessment can help the IT staff, system administrators, and the C-Level Executives of a business or organization determine if their security needs can be met with just a biometric solution, or if another biometric or even non-biometric technology is needed to create a multi-modal security solution, in order to fulfill the specific security requirements.

## System requirements documentation

With biometric requirements documentation, this is where all of the steps and components of the proposed biometric system are documented and tracked. After the biometric system is implemented, this documentation then serves as the focal point for troubleshooting and creating end user training booklets and guides.

In terms of biometric system requirements validation, this is the process where all of the stakeholders who are involved in the implementation review the documentation created in the previous step to find any errors, or points of confusion in order to ensure that the biometric system implementation goes as smoothly as possible, and that all foreseeable problems are resolved ahead of time.

Also, prototypes of the actual biometric system can be obtained from the vendor and a test environment can be simulated to further ensure that the real biometric system implementation will go smoothly, and just as planned. Once all of this has been accomplished, then all of the stakeholders sign off for the deployment of the biometric system.

## System requirements validation

In the last step, the biometric requirement management phase, new documentation is created and updated so that it remains valid if upgrades and/or changes occur in the newly deployed biometric system. Also, during this entire process just described, the place of business or organization can put out RFPs to various biometric vendors, to help narrow down which biometric systems will fit into the budget and meet the security requirements.

Also during this timeframe, the biometric vendors who make the shortlist can also provide to the business or organization trial products to closely evaluate which biometric system will meet the security needs the best.

## Biometric system specifications

After the biometric system requirements analysis phase, the next phase in the biometric project management lifecycle is that of the biometrics system specification phase. This part of the project management phase provides for what the biometric system will accomplish and what it will not. This phase helps the C-Level Executives keep the biometric system costs in check and keep the implementation time running as scheduled.

The biometric system specification phase consists of the following components:

1. Biometric system functionality: These are the specific security tasks that the biometric system is supposed to accomplish;
2. Biometric system external interfaces: These are the specifications for the biometric system will interact with, such as the software applications and the other, related hardware applications;
3. Biometric system performance: These are the specifications as to how each subcomponent and the entire biometric system as a whole is expected to perform;
4. Biometric attributes: These are primarily the specifications for the biometric system maintenance;
5. Biometric system design constraints: These are the best practices and standards the newly implemented biometric is supposed to follow;
6. Biometric system testing: These specifications lay out the criteria that will confirm what is acceptable performance for the biometric system.

From these above components, the biometric system specifications can be written internally by the project management team from within the business or the organization. Or, it can also be written by a neutral, unbiased third party as well. But regardless of who actually is composing and writing the documentation, the biometric system specification document will consist of the following components:

1. The introduction: This portion of the document consists of the executive summary, the scope of the biometric system deployment and implementation, key terminology used in the process, and references used for research conducted to justify the need for a biometric system;
2. Product description: This portion describes the entire biometric technology that is going to be used at the place of business or organization, and how it will be deployed, for example:
    a. The total number of biometric devices to be deployed both outside and inside, as well as the total number of end users expected to use each biometric device;
    b. Who will be made members of the biometric system administration and support staff;
    c. The envisioned biometric system networking environment (which will be described in detail in the next section);
    d. How training will be conducted for the end users;

   e. How the privacy of the end user's biometric information and data will be enforced.
3. The system requirements: This final portion of the document lays out the details of the formal security requirements for the biometric system to be procured and deployed.

## System architectural and processing designs

The next phase in the biometrics project management phase is that of the biometric system architectural and processing design. This is the blueprint that is utilized to specify the very exact biometric hardware and software that will be deployed at the place of business or organization. In terms of biometrics, the architectural design component consists of three primary objectives, which are as follows:

1. High-level goals: This describes the final biometric system security goals;
2. Mid-level objectives: This describes where the biometric system will be placed, and how the biometric templates will be matched with, and compared with another;
3. Detail high-level objectives: This helps to define which specific hardware and software will be used in the biometric system.

In other words, whatever the previous phase (the systems specifications phase) examined, it is all now decided upon and finalized by the stakeholders in this phase. This phase consists of two primary objectives, namely, the functional architecture and the operational architecture. These two components also consist of subcomponents and are described as follows:

1. Functional architecture: This component describes the actual physical aspect of the biometric system hardware and the type and kind of environment it will be placed in at the business or organization. It addresses the following subcomponents:

   a. Environment design: This describes how the ergonomic environment will be designed for the biometric system;
   b. Hardware design: The specific design of the hardware of the biometric system is a direct function of the exact purpose of the biometric system, and the type and kind of environment it will be placed in. The hardware design of the biometric system is also strongly correlated with the operational architecture, which is discussed next. It is also important to note that the biometric system hardware design is also strongly correlated to the existing security requirements of the business or organization,

as well as the existing security infrastructure. As a result, the proposed biometric system may be able to integrate with legacy security systems, thus providing for multiple layers of security defenses.

2. Operational architecture: This type of architecture lays out the design of the software system of the entire biometric system, and this is what enables the functional architecture to do its part of the work as well. It should be noted that this design of the software system, at this point, is only high level. It does not go into the granular level of detail as to how the exact software code should be written, but there will be enough detail in it so that the software developers who are involved will know at a cursory glance what exactly needs to be developed. Within this architecture component, the goals of determining how the biometric templates will be stored, and where the matching algorithms will be stored are also determined. In terms of multimodal biometric solutions, the operational architecture also specifies how the different biometric modalities will work together.

## Storage and matching combinations

From this, six different types of storage and matching combinations are most commonly used, and they are as follows:

1. Store on server, match on server architecture: In this type of architecture, the biometric raw images that are collected from the sensors, and the subsequent templates that are created are both stored and compared/matched with one another at the server level. This is also one of the most utilized architectures, and the typical application for which this is used is large-scale (1:N) scenarios;
2. Store on client, match on client architecture: With this architecture scenario, the biometric raw images and the biometric templates are stored at the client level, and all processing of the data and information is conducted at the local level. This type of architecture can operate and function without a centralized network, and the most common application for this particular architecture is that of single-sign solutions for smartphones and laptops;
3. Store on the device, match on device architecture: In terms of this operational architecture, the biometric raw images, and the biometric raw templates are stored upon the actual authentication device itself, and again, all processing of information of the data and matching of the biometric templates takes place at the local level. The most common application for this is for verification scenarios on wireless devices;

4. Store on token, match on token architecture: Under this operational architecture regime, the biometric raw images and the corresponding biometric templates are stored on some physical token, such as a smart card. The processing of the biometric templates takes place over a wireless transmission between the smart card and the smart card reader. Because the information and the data are stored on the smart card, different biometric modality templates can be stored on it as well. The most common application for this particular architecture is that of credential-based security systems, such as turnstiles for physical access entry at large office buildings;
5. Store on the token, match on device architecture: In regards to this type of operational architecture, and just like the above-described architecture, the biometric raw images, and the corresponding templates are stored onto the memory chip of the smart card. But the primary difference is that the matching and the processing all take place on the smart card itself, without the need for a wireless connection and a smart card reader. One of the most common applications for this type of architecture is that of worker-based credentialing systems;
6. Store on token, match on server architecture: With this type of operational architecture, the biometric raw image and the corresponding templates are stored onto a smart card, but the processing and the matching of the biometric templates take place at the server level. A primary advantage of this is that there is no central storage of the biometric templates involved, and all decision-making occurs in a very secure environment. A very common application for this kind of operational architecture is that of network and internet access.

If one were to actually compare all of the subsystem locations just reviewed, a cross-comparison of these scenarios can now be made. For example, storing biometric templates at a server level could prove to be much more complex and much more financially expensive than a local storage device.

For instance, if the server level is utilized, there will be costs associated with licensing, hardware and software upgrades, as well as hiring a full-time server administrator. Therefore, the use of a server-based system would probably work best for large-scale identification-based applications. For verification-based applications, local storage, such as the level of a biometric device or a smart card, possesses some unique advantages, such as:

1. Other types of wireless readers can be used (such as for the e-Passport applications);
2. No need to have the costs associated with a centralized server and database;
3. Multipurpose biometric applications can be much easily deployed and served to the end user base.

## Operational architecture design-multimodal systems

Point 3 above leads into the operational architecture design for multi-modal biometric systems. It is important to keep in mind that multi-modal biometric systems compose an entire security package together, therefore, different types and kinds of biometric information and data will be shared. This is also referred to as fusion, and there are four distinct levels of it, which are as follows:

1. Sensor level fusion: At this particular level, biometric information is the most unique and rich, but also contains a lot of extraneous background noise which could greatly affect the template creation process;
2. Feature level fusion: Biometric information and data at this level are much less unique than the previous level, but also, the extraneous noise is much less. There are also two different sub-fusion layers at this particular level, which are as follows:
    a. Unimodal feature fusion: Single biometric information and data fusion can be applied;
    b. Feature normalization: A biometric system with different modalities may require normalization, which extracts unique features based on statistical means and variances.
3. Score level fusion: At this stage, biometric information and data are extremely simplified, and this is a very common method that is used the most today, because of its ease of use;
4. Decision level fusion: At this particular level, the biometric information and data are very much limited, and this technique is only used when no other biometric information and data is available. This is also referred to as rank-level fusion in large-scale identification-based applications.

A basic principle to be remembered here is that when there are multiple sources of biometric information and data which is available, this is obviously much better in terms of unique feature extraction rather than just relying upon one source of biometric information and data.

## The information processing architecture

Within this also lies what is known as the information processing architecture. This typically involves the design of any type or kind of computer that is needed, as well as the mathematical algorithms and the computer networks needed to support the entire biometric system at the place of business or organization. These are reviewed in the following order:

1. Computer hardware: For biometric systems, speed of course is one of the most important functions. But it is not the speed of the biometric system that is important, but it is also the speed of the computer that is equally important. After all, as discussed previously, if a large-scale biometrics system is deployed for identification, obviously a centralized computer server is needed, not only to store the enrollment and verification templates but also to process all of the transactions that are required. Because all of this requires extra resources, speed will be required on the part of the server processor. In this regard, speed is a direct function of:
   a. The generation of the CPU, and its speed is measured in Hertz;
   b. The arithmetic logic unit which regulates mathematical calculations;
   c. The bus speed of the CPU, which is measured in the speed and the width;
   d. The input/output rate at which biometric information and data can be sent to other hardware devices which support the entire biometric system.

Parallel and distributed processing: In most large-scale biometric systems, just one processor is not enough. Therefore, the workload must be shared. In order to accomplish this, parallel and distributed processing is often utilized, and is broken as follows:

Parallel processing: This is defined as "the simultaneous transfer, occurrence, or processing of the individual parts of the whole, such as the bits of a character, using separate facilities for the various parts" (Source: Certified Biometrics Learning System, Module 2, Biometrics System Design & Performance, Copyright 2010 IEEE, pp. 3–90).

In other words, one server processor can send the workload into different components to other server processors to execute a portion of the larger program which supports the biometric software applications from within the system;

Distributed processing: With this type of processing, the same task is broken amongst a series of server processors. In this instance, a single mathematical instruction, and multiple data architectures are utilized. This simply means that the server processors perform the same, repetitive tasks on different sets of biometric information and data.

The parallel and distributed processing components as described up above are a direct function of what is known as "execution efficiency". It is specifically defined as "the degree to which a system or component performs its designated functions with minimum consumption of time" (Source: Certified Biometrics Learning System, Module 2, Biometrics System Design & Performance, Copyright 2010 IEEE, pp. 3–32). This can also be referred to as what is known as "computational expense".

This is greatly impacted by the sheer complexity of the mathematical algorithms that are involved in a biometric system, as well as the sheer amount of biometric information and data that must be processed. As a result, it should be noted that there is a distinct tradeoff between the processing quality of the biometric templates and the amount of memory that is required.

## Subsystem analysis and design

The next phase of the biometrics project management lifecycle is that of the biometrics subsystem design phase. In any biometric subsystem, in theory at least, the subsystems are designed to be logically separated from one another. In order to accomplish this task, the concepts of coupling and cohesion are utilized. Coupling can be defined as "the manner and degree of interdependence between software modules"; and cohesion can be defined as "the manner and degree to which the tasks performed by a single software module are related to one another" (Source: Certified Biometrics Learning System, Module 2, Biometrics System Design & Performance, Copyright 2010 IEEE, pp. 3–92).

For example, coupling is used and designed in such a way that each subsystem in the overall biometric system can operate separately from one another. To illustrate this point, as the biometric information and data are passed from one subsystem to the next, control over each subsystem is also maintained in terms of execution.

With respect to cohesion, every subsystem in the overall biometric system only performs and executes those specific tasks it has been assigned, in order to ensure that the biometric system works as one harmonious unit.

In this part of the biometric subsystem, it is important to keep in mind that the biometric system is intended to be what is known as the "human to machine interface". As a result, the choice of sensor becomes of paramount importance, in order to capture and collect the biometric raw images (in this respect, physical biometrics, primarily).

The design and ultimate choice of a sensor not only has a huge and direct impact on the performance of the biometric system, but a phenomenon known as "sensor fatigue" can also set in over the useful lifetime of the biometric device. This occurs when the sensor can no longer capture and collect any useful and meaningful biometric raw images.

Also, the environment in which the biometric system will be placed and the type of application it will be used for are also other important considerations that need to be taken into account, and this can have a direct impact on the choice and design of the sensor as well. The signal and image processing subcomponent of the overall biometric subsystem is dependent upon a number of key factors, which include the following aspects:

1. The specific biometric modality which will be required by the place of business or organization;
2. The type of sensor which is needed;
3. The preprocessing and the feature requirement of the biometric raw image;
4. Any changes to the preprocessing order;
5. Any modular designs which can be incorporated;
6. The types and kinds of compression ratios which are required for processing the biometric raw images.

## Data storage subsystem design considerations

Next in line after the signal and image processing is that of the data storage subsystem design. As discussed before, this is the subsystem which consists of the database which houses the biometric templates, and the transaction processing which occurs.

Biometric database design is a very complex task, and at a macro level, and some of the factors such as the overall biometric system requirements, and what the exact security needs of the business or organization need to be examined closely. Also, another key element which helps to determine the biometrics database design is the overall management of the database.

For example, small database designed for verification (1:1) applications will be much easier to manage than a large-scale database application (1:N). Also, the database designer(s) need to decide if the biometrics database will only contain the enrollment and the verification templates, or if other types and kinds of biometric information and data will be incorporated into the database or not.

A smaller database of course means less processing power is required, and the search time is greatly enhanced. Another issue involved with the biometrics database is design is where to store the biometric raw images. Typically, these raw images are not stored in the biometrics database, but if they are stored for any type or kind of reason, the following factors must be taken into consideration:

1. The biometric raw images should be kept in a separate database in order to optimize performance;
2. The search time for the biometric raw images can be much slower because they are not accessed frequently;
3. The biometric raw images, if they are stored in the database, will need to be reformatted, so that they can be used to replace the biometric templates which degrade in quality over time, or accidentally deleted from the system;

4. The biometric raw images need extra security layers, but these protocols should not be applied to the database which houses the enrollment and verification templates, as it will greatly lessen the search and retrieval time.

## The DBMS

A biometrics database is normally, run, administered, and maintained by a database management system, also known as a "DBMS". A place of business or organization can either utilize a closed-source database, or it may utilize an open-source database. Both of these have advantages and disadvantages. For example, with the former, although there would be strong vendor support, the software and the respective licensing fees would be cost prohibitive for a small to medium sized business.

But with the latter DBMS option, the source code is virtually free, but due to its open-source nature, it can be prone to large-scale attacks and other such vulnerabilities. The DBMS for a biometrics database can either be small or large. If it is large, then obviously other considerations need to be taken into account such as:

1. The maximum number of end users and their corresponding templates which can be stored into the biometrics database;
2. How the biometrics DBMS will handle search and retrieval requests;
3. The overall design of the DBMS, specifically the database structure;
4. If and when parallel and distributed processing (as discussed previously) will be utilized;
5. Specific disaster recovery plans for the DBMS and the corresponding biometrics database structure;
6. The frequency in which the biometric information and data in the DBMS will be "cleansed".

The matching subsystem (where the enrollment and verification templates will be matched against) design will be dictated primarily by the type of biometric modality and the vendor who will be ultimately decided upon by the business or organization. Every system design tries to have a compact matching subsystem component, but this of course is 100% dependent upon where the matching will take place.

If the matching subsystem will be placed in a client–server network topology, the matching subsystem can be designed as small as possible, because the server has dedicated resources which can be used. But, if the matching subsystem is to be placed onto a smart card, then the resources for matching have to be as maximum as possible, given the constrained nature of a smart card.

With regards to the decision subsystem design (this is the part when the final decision is made whether or not to confirm the identity of the person in question, and whether or not to grant them access to the resources they are seeking to use), other considerations need to be taken into account as well, such as:

1. Whether a supervised or a non supervised system will be utilized;
2. The point at which positive verification or negative verification (similarly, also positive verification and negative verification) will be made based upon the comparison of the enrollment and verification templates;
3. If only one security threshold will be used, or other security thresholds will be used as well in order to provide a sense of randomness to the biometric system.

## Determining the security threshold

As it was reviewed earlier in this book, deciding upon where the exact security threshold be set at can be a very complex process to achieve, in order to garner the most optimal result from the place of business or organization. For example:

1. Very often, ascertaining the exact level of the security threshold is first done in a test environment. If the end user database varies from the test database, then the ascertained security threshold setting will be completely offbase;
2. A very strong barrier in ascertaining the true, optimal security threshold setting is further exacerbated by the fact that it is statistically very difficult to determine what the true rejection rates are.

Also, two other key factors which are very important in ascertaining the security threshold level are the type of biometric modality which is being used, and the exact application environment it will be placed in. But, if a variable threshold is decided upon, then different weights of importance will have to be assigned to the numerous criteria which are being utilized.

Finally, the C-Level Executive should keep in mind that there is a steep trade-off between end user convenience (which means having a very low Failure to Accept Rate) and the level of security (which means having a very low False Acceptance Rate) in either both verification or identification applications, with regards to implementing the optimal security threshold setting.

For example, if the convenience of end users is valued over security, this is called a liberal security threshold setting. But, on the opposite realm, if security is valued over end user convenience, this is then called a conservative security threshold setting. Most biometric applications lie in between these two extremes.

With regards to the administration subsystem design, this is where the biometrics system administrator has all of their controls located at, via the control panel. The design of it once again is heavily dependent upon the type of biometric system which is being used, and the particular vendor who designed it.

Finally, in terms of the data transmission subsystem design, this is essentially the networking component of the biometric system, and will be discussed in greater detail in the next section. Generally speaking, the simpler the data transmission design is, a higher level of security can be offered.

But on the converse, the more complex a data transmission design is, this usually translates into greater points of system failure as well as potential security breaches. A very key, but often neglected component in subsystem design is that of the biometric interchange formats, which spells out how the biometric templates will traverse safely across the network medium in the data transmission subsystem.

Before any biometric system can be fully deployed and declared to be operational by the place of business or organization, the system must first be fully tested in order to fully ensure that it fully meets the security requirements, and most importantly, that all of the technologies involved are properly functioning in the way that they should be.

## Subsystem implementation and testing

And in fact, this is the very next step into the biometrics project management lifecycle: That of the biometric system implementation and testing. The testing of an actual biometric system can be broken down into three distinct parts:

1. Biometric subsystem level testing: This is when the biometric devices in the entire system to be deployed are tested first each individually, and in a stand-alone mode;
2. Biometric system integration level testing: This occurs when the individual biometric devices in the aggregate system are all joined together into their respective systems and tested as different groups in order to make sure that each and every biometric device within their groups are all working harmoniously together;
3. Biometric system wide level testing: This is when the biometric system is tested as an entire whole, or one cohesive security unit, in order to fully ensure and guarantee 100% that everything is working perfectly.

Throughout the three phases of the biometric system as described up above, a key aspect which is of prime importance is that of the quality assurance checks which have to take place. The quality assurance lifecycle ensures that all of the bugs and errors found during the system testing have been fully corrected, and the entire biometric system functions fully at the level it was designed and intended to do.

The quality assurance lifecycle includes the following:

1. Verification: This is the quality assurance process which ensures that the newly implemented biometric system is totally free of errors in the hardware and the software;
2. Validation: This is the quality assurance check which helps to guarantee that the results of the newly implemented biometric system matches up to the original project management plan.

Another type of test, which is optional, is that of the biometric system prototype testing. This is when a smaller version of the entire biometric system, or if you will, a snapshot of it is implemented at the place of business or organization. The goal of this kind of test is to see how a small, but representative sample of the biometric system will work in a live environment.

The goal of biometric system prototype testing is to fully understand the risks, and the type and kind of end user training which will be required before considering a full fledged biometric system implementation. In other words, this type of testing allows for a business or organization to take certain and calculated "safe" risks to see if a biometric technology will work best to meet their particular security needs and requirements.

Finally, the very last type of test which is required is that of biometric system acceptance testing. This is one of the most important phases of the biometric project management lifecycle, as this proves that the actual biometric system is working, and fully functional. Under this testing, a representative sample of the end user group is also asked to participate, and if all goes well, the final and sometimes ceremonious sign off occurs, and the biometric system then goes 100% live and operational at the place of business or organization.

## System deployment and integration

Another key aspect in the biometrics project management lifecycle is that of biometric system deployment and integration. With this component, the place of business or organization contemplates whether or not implement a biometric system as a security add on to their existing security infrastructure, in order create a multimodal, or a multiple security defense layer.

The key question which needs to be answered here is if the biometric system will be the prime means of security for the business or organization, or if the combined security system (which is the existing security system plus the new biometric system) will provide the total security protection.

Very often though, it is the latter security model which is typically chosen. In this case, it is up to the C-Level Executives and their respective IT staff to figure out how the new biometric system will integrate with the existing security infrastructure. For this big task to be accomplished, a number of key questions need to be answered such as:

1. Just exactly at what points will the new biometric system interface with the existing biometric system?
2. Will the newly implemented biometric system take second priority over the existing security infrastructure? In other words, will the new biometric system be considered just as an add on to the existing security infrastructure?

In this situation also, the new biometric system becomes a "technical interface" to the existing security infrastructure system. It can take this shape in three different ways:

1. The biometric system can exist as a tightly woven interface with the existing security system;
2. The biometric system can exist a loosely woven interface with the existing security system;
3. The biometric system can act as a middleware interface with the existing security system.

With a loosely woven interface, this is merely a simple matchup of the outputs of the existing security system with the inputs of that of the biometric system. This type of communication is very often as implementing a networking protocol between the two systems.

In terms of a tightly woven interface, the C-Level Executives and the IT staff, in this case, will decide how the new biometric system will fit into and work seamlessly into the existing security infrastructure. The resultant of this is a multimodal security system which operated as one, cohesive unit. However, it should be noted that the new biometric system just won't fit nicely and easily into the existing security system, modifications will have to be made to the latter in order to accommodate the former.

Finally, with regards to the middleware interface, this is a biometrics management system which is literally sandwiched between the two layers of the existing security system. Middleware can be considered a watered down version of a full fledged biometric device, and it has its own functionalities as well.

## Middleware

Biometrics middleware can also be considered to be a cost alternative to a full blown biometrics system. Biometrics middleware serves the following functions:

1. Act as an integrating device between the various legacy systems;
2. Have a free flow of communication with the various biometric systems (such as optical-based, ultrasound-based, and silicon-based);
3. Help to manage the various biometric databases;
4. Work as a liaison with the networking protocols in the existing security infrastructure;
5. To work in an open-source security hardware and software environment.

It should be noted at this point that biometrics-based middleware is either client side or server side. For a example, with a client side interface, the biometrics middleware can support a whole host of biometric sensors from all types and kinds of vendors.

With a server side interface, the biometrics middleware assumes all of the backend management of the entire security system, which involves verification and/or identification transaction processing. If a biometrics system is deemed to be very large in nature, than a biometrics middleware known as a "enterprise solutions middleware" is often required.

This type of biometrics middleware solution is both either client side or server side, or even both at the same time. This type of middleware solution is designed to handle much more complex tasks and functions, such as that of large-scale identification applications.

## The biometric interface

The interface to the new biometric system and the existing security system can be prone to the security breaches from both outside and inside the place of business or organization, and therefore, the appropriate internal controls need to be applied in order to mitigate such risks.

The interface can either be a standard out of the box solution, or it can be designed entirely from scratch in order to meet the exact needs of the business or organization. In this case, many customized subcomponents will be required such as a programming language translator (in order to ensure the communication between the new biometric system and the existing biometric system is that of a common protocol), as well as highly specialized security protocols will be needed as well, such as high-level cryptography for the scrambling and descrambling of the biometric information and data, as well as the other types of and kinds of security data from the existing security system.

The actual biometric system implementation involves the actual installation of the biometric devices and the related networking components, and finally training the end users in how to properly and accurately use the biometric system.

An often forgot about obstacle of implementing an actual, live biometric system is that of the logistics which will be involved and required. For instance, dedicated facilities may be required in order to properly house the biometric system (of course though, depending upon what the specific nature of the application is), and physical changes to the building which houses the business or organization may required as well.

## End user/administrator training

Along with end user training, just as equally important is that of the biometric system administrator training as well, which consists of the following:

1. An understanding of the specific operation of the entire biometrics system;
2. Training in the proper maintenance of the biometric devices;
3. Education on how to actually train in the end users (this is very often the case if the biometrics vendor is not conducting the actual, end user training).

## System maintenance

The very last step in the biometric system project management lifecycle is that of the of the biometric system maintenance. In simple terms, these are the steps which are required to keep the biometric system running in peak order and optimal conditions. Very often, one thinks up of cleaning and system tweaking of the biometric hardware in terms of this maintenance component.

But, even collecting the data which is recorded by the biometric system can also be used to keep the biometric system in optimal working order. Such data can be collected from the biometric system logs and transactions reports. For instance, the following types of data are useful in keeping a biometric system in good, working order:

1. The statistics related to the total number of users which have been accepted and rejected; the total number of FTE rates; and also eh total number of attempts which are required by the end user before they are ultimately accepted or rejected by the biometric system;
2. The actual number of FMNRs recorded by each biometric device in the entire biometric system;
3. The actual FMRs recorded by each biometric device in the entire biometric system.

As mentioned just previously, a big component of biometric system maintenance is fine tuning, which includes the following duties:

1. Manually adjusting the security threshold levels;
2. Implementing additional end user training when needed;
3. Calibrating the biometric hardware devices within the entire system;
4. Changing or reconfiguring the software which drives the applications of the entire biometric system;
5. Tweaking the network performance and bandwidth between the biometric devices and the central server;
6. Keeping the mathematical algorithms (especially that of the matching algorithms) totally optimized with the latest upgrades and developments form the biometrics vendor;
7. Keeping the security threshold level just at the right point so that the security needs of the business or organization will be perfectly met.

## Upgrading an existing system/fine tuning

Also, just like software applications, biometrics technology too, has their fair share of releases and version numbers as well. These version number releases can come out for an entire biometric system, or for just the individual devices which make it up.

An important consideration to take into account when upgrading to a newer biometric system version is that of template aging. This refers to the degree of dissimilarity between the enrollment template and the end user's actual physiological or biological traits which have changed over time.

This could be caused by the aging process which we all experience, or any weight changes, or even any injuries. A perfect example of this is that of facial recognition. Over time, a person can go through massive weight loss or massive weight gain, thus rendering the initial enrollment templates as useless.

As a result, the facial recognition system will not recognize the particular individual, thus causing he or she to go through the enrollment process all over again. It should be noted that all physical and behavioral biometric technologies are not prone to template aging.

A good example of this is that of iris and retinal recognition. As reviewed previously in Chapter 2, the biological and physiological structure of the iris and the retina hardly ever changes over the lifetime of the individual. But, if the technology in the biometrics system is advanced enough, for example, if it contains neural network technology, it can create a replication of the original enrollment template, using various modeling techniques.

This will prevent the end user from having to re-enroll their templates again after a period of time. This technique is known as "template/model adaptation". In the biometric system, it could also be the case that the biometric templates could be set to expire at a certain date, thus rendering the need for re-enrollment by the end users.

Typically, biometric templates are not usually set to expire, but for some reason or another, the system administrator may establish this particular requirement. The method of expiring biometric templates involves the deleting the enrollment templates all together from the biometrics database, and having the end users re-enroll fresh templates all over again.

## System reports and logs

Finally, as mentioned through the previous chapter of this book, the reports and logs which can be outputted from a biometric system are an important tool for the system administrator to analyze. These types of reports and logs can consist of the following information and data, depending upon the vendor who manufactured the particular technology:

1. End user names;
2. Details of the verification/identification transactions and their corresponding data;
3. The decision (if authorization or non-authorization was the end result);
4. The security threshold setting which was applied to the decision made in 3.

The following are the benefits of the reports and logs which are outputted by a biometric system:

1. Monitoring the performance of the biometric system and the devices which comprise it;
2. Keeping track of the version history of the modifications and changes made to the database;
3. Creating a repository of all of the data associated with the biometric system (such as performance metrics and statistics) for later analyses;
4. Providing a paper based audit trail of the end users utilizing the biometric system.

It should be noted that these reports and logs from the biometric system are designed to assist with the further optimizing of the biometric system to reach its peak levels, and to trouble shoot any types or kinds of problems.

## System networking

Another very important component to the overall biometric system architecture and the biometric project management lifecycle is that of the networking systems which can be applied. As it was eluded to earlier in this book, when a C-Level Executive does indeed think about a biometric system, the image of a standalone fingerprint scanner, or that even of an optical scanner attached to a laptop or netbook often comes to mind. But what often gets overlooked, especially by the C-Level Executives is that an overall biometric system can be indeed composed of many other different biometric modalities, and all of them can be networked either physically or via a wireless connection with each other.

In fact, many biometric systems today consist of not just one biometric device, but they consist of many devices which are all interlinked together. Because of this interlinking with another, the biometric devices can literally talk with another, share templates, and also even transmit other types and kinds of biometric information and data.

## Network processing loads

In fact, when the biometric devices are all connected with another, the transaction processing load for verification and identification applications can be shared, with distributed network resources. A perfect example of this is that of an airport application. Today, facial recognition and iris recognition are the primary two biometric modalities which are being used to track and identify the passengers at the major international airports, all over the world.

Because of the sheer bulk of the number of passengers and travelers coming in and out of these airports, it is impossible for just one facial recognition device or even just one iris recognition device to do all of the verification and identification. Therefore, many devices are placed at various, strategic locations throughout these airports, and are all networked together in order to ensure that all of the transaction processing occurs between all of the devices.

Also, another key advantage with networking the biometric devices together in such a large environment is that if a potential terror suspect is lurking throughout one of these major international airports, an elude surveillance by some of the biometric devices, the chances are high that they will be identified by some of the other biometric devices, and subsequently apprehended.

This all due to the fact of the networking of all these different biometric modalities together, and thus also, possessing the ability to share critical information and data with one another. This airport example is just one type of networking scenario.

## A networking scenario

The other major type of networking scenario which can take place is when all of the biometric devices can be still networked together, but in turn, all of these biometric modalities in the overall system design are also connected to a central server (and perhaps even more that one server can be utilized, such as the large AFIS fingerprint biometric databases administered by the FBI, where all of the processing of verification and identification applications take place.

A good example of this is that of physical access entry. Imagine if you will, a large-scale business, such as that of a warehouse with thousands of employees. Suppose also that fingerprint scanners are installed all over the factory space. Obviously, for one fingerprint scanner to do all of the work of confirming each and every employee would be way too much (not only that, but also very time consuming as well).

But, will of the fingerprint devices networked to a central server, all of the transaction processing can take place at this large-scale very quickly and easily, since the sheer computing power and the resources the server possesses is much greater than just one fingerprint scanner.

## Biometric networking topologies

Both of these types of networking topologies will be reviewed in greater detail later, in this section. As a result, it is very important for the C-Level Executives and their IT staff to have broad understanding of the various networking topologies that a biometric system can take up, and how that networking will interface with their existing security system.

Thus, the goal of this section of this chapter is to provide the C-Level Executive with a basic understanding of the networking concepts and designs that a biometric system can take on, and to provide a primer when a biometric vendor discusses about the networking their particular biometric solution can work with. But, it is important for the C-Level Executive to keep in mind that the specific design of the networking structure of a proposed biometric system will be heavily dependent upon the new security requirements, as it is dictated by the needs of the business or organization.

This section will review the following:

1. Data Packets;
2. Data Packet Switching;
3. Network Protocols and TCP/IP;
4. Network Topologies including Peer to Peer and Client–Server Architectures;
5. Routers and Routing Tables.

## Data packets

In the world of networking, and even including biometrics, one of the most fundamental units of connectivity is that of the data packet. For example, every time that an e-mail is sent, or a biometric template is sent to a central server, this is all done via the data packet.

In other words, for two biometric devices to communicate with the central server, or even with each other, the data must first be broken down into smaller bits, travel across the network medium, and upon reaching the central server or the destination biometric device, the data must be reassembled again into a format that the central server or the destination biometric device can decipher and comprehend.

For instance, a regular size biometric template would be simply too big in terms of size to traverse quickly and easily across the network medium. In order to accomplish this task, the biometric template must be broken down into very small bits by sending biometric device, and it is these small bits which are known as the data packets.

These data packets (the broken up biometric template) then travel across the network, then once they reach the destination biometric device, the data packets are then reassembled again into its whole form and entirety, so that the destination biometric device can understand the data it has received, and take the appropriate actions on it which are necessary.

A definition of a data packet is:

> a basic unit of communications [or data] over a digital network. A packet is also called a datagram, a segment, a block, a cell, or a frame, depending on the protocol. When data has to be transmitted, it is broken down into similar structures of data, which are then reassembled to the original data chunk once they reach their destination.
>
> *Source:* http://en.wikipedia.org/wiki/Network_packet

To further elaborate upon this definition, a data packet consists of a header, the payload, and the trailer. The header specifies the destination biometric device (or central server) where the data packet has to go, the payload section consist of the broken down biometric template, and the trailer possesses the ability via a mathematical algorithm, to make sure that the data packet gets to the right destination biometric device, or the right central server.

## Data packet subcomponents

In technical terms, the data packet is broken down into various subcomponents, which are described as follows:

1. A total of 4 bits which specify the version type of the network data packet protocol the biometric template resides in. At the present time, there are two types of network protocol data packets, which are known as IPv4 and IPv6;

2. A total of 4 bits contain what is known as the Internet header length, and this specifies exactly how long the header of the data packet is which contains the biometric template;
3. A total of 8 bits contain the Type of Service, also known more commonly as the Quality of Service. This merely refers to what priority level the data packet will have in terms of precedence upon reaching the central server, or the destination biometric device. These bits also dictates, to a very small degree, which biometric template has what priority in terms of verification and identification processing;
4. A total of 16 bits which actually contains the length of the data packet by itself. The specific size will vary greatly upon the size of the biometric template;
5. A total of 8 bits which specifies the Exact Time to Live, and this is the number of hops that the data packet is allowed to take between servers or biometric devices before it is expired and discarded;
6. A total of 8 bits which specifies the network protocol that the data packet consisting of the biometric template will traverse across (most likely it will be the TCP/IP protocol, and this will be reviewed in greater detail);
7. A total of 16 bits which possess the Header Checksum, and this contains a mathematical algorithm (specifically, the Cyclical Redundancy Checking (CRC) algorithm). This is used to check for the biometric template integrity once it reaches the central server or the destination biometric device. The CRC algorithm also detects for any errors which may have occurred during the transmission of the data packet;
8. A total of 32 bits which contain the Source IP Address form which the data packet originates from (in this case, it will always be the sending biometric device);
9. A total of 32 bits which contains the destination address (in this case, this would be the destination biometric device, or the central server).

## Data packet switching

A very important aspect of biometric network design is that what is known as data packet switching. For example, when at least three or more biometric devices are connected together, or when a biometric device is connected to three or more central servers, different network routes may be used to reach the destination biometric device or the central server.

This is done to optimize the flow of data packets in the overall biometric network system design. A key question which now often gets asked by C-Level Executives is what makes the connection possible between two or more biometric devices, or two or more biometric devices which are connected to a central server?

## Network protocols

In other words, they are all networked together, but what makes it possible for the data packets containing the biometric templates actually flow freely between the biometric devices and the central servers? The answer lies in what is known as the network protocol. Specifically, a network protocol can be defined as "the format and the order of messages exchanged between two or more communicating entities, as well as the actions taken on the transmission and/or receipt of a message or other event" (Source: *Computer Networking: A Top Down Approach*, Kurose, J.F. and Ross, K.W., 2008, Pearson Education, p. 8).

Today, all biometric devices connected with each other have to make use of a certain network protocols in order to transmit their data packets back and forth. There are many network protocols which are in existence today. These include very specific protocols for open-sourced applications, closed-source applications, hardwired connections (this simply means that the two biometric devices are connected by an actual, physical cable), and wireless connections.

## TCP/IP

But in the world of biometric system architectures, the network protocol known as TCP/IP is what is used most commonly. In fact, this particular network protocol is also what is used in most applications which relate to the Internet. TCP/IP is actually a combination of two other network protocols.

TCP stands for Transmission Control Protocol, and IP stands for Internet Protocol. This combination protocol of TCP/IP is a very powerful one, and in fact, it goes as far back as the 1970's, when it was first conceived of when the very first, prototype Internet came out, known as the APRANET.

TCP/IP consists of four distinct network layers, which are further defined as:

1. Link layer: This layer consists of the communication tools used in the biometric network so that the data packets which contain the biometric templates can travel from one link to the next link until it reaches the destination biometric device or the central server;
2. The Internet Protocol: This is the IP component of the TCP/IP protocol, and is primarily responsible for the internal communications between the various network links which make up the entire network segments between the biometric devices, or the network segment between the biometric devices and the central server;
3. The Transport Layer: This is the TCP component of the TCP/IP protocol, and is primarily responsible for the overall communications

between the biometric devices, or the biometric devices and the central server;

4. The Application Layer: This is where the overall TCP/IP protocol resides at, and is responsible for all of the processing services which take place, such as the verification/identification transaction processing, or the template matching processing.

As it has been mentioned before, at the beginning of this networking section, there are two main types of network designs which are possible. For example, biometric devices can be networked to one another, or the biometric devices can be all networked to a central server.

## Client–server network topology

Now, we give these two types of network designs specific names. The former can be described as a peer to peer biometric network, and the latter can be referred to as a client–server biometric network. In the client–server biometric network, the server is known as the host, and the biometric devices which connect to it are called the clients. For an example of a client–server network topology, see in Figure 3.3.

The clients in this type of network architecture merely transmit the biometric templates and other relevant information and data to the host. It is from here that the host then conducts and processes the verification and identification transactions, and also contains the databases which house both the enrollment and verification templates.

The host is always on, and the client can either be on or off, depending primarily upon what time during the day at which they are needed. The transmission of the biometric templates to the host are known as network requests (in other words, the client is asking the host to confirm the identity of the individual in question, and to send the result of this transaction back to the client).

Another notable feature of the client–server network biometric topology is that the clients never actually communicate with another. Rather, they only communicate with the host, which services the network requests. This is so because the host has a unique TCP/IP number attached to it. Based upon this, the clients know which host they need to send their network requests to.

Of course, depending upon the scale and the magnitude of the client–server biometric network topology, often just using one host is not enough to process all of the verification and identification transactions, and to store all of the biometric templates in.

Very often, two or more hosts are needed, and depending upon the budget of the business or organization, and the total number of end users involved,

*Figure 3.3* This is an image of client–server network topology. (Black and gray amplifier beside white and black surface photo – free server image on Unsplash.)

a server farm may be required. In these instances, many hosts (servers) will be required.

## Peer to peer network topology

In the peer to peer biometric network, there are no main hosts which are required. In this situation, as described before, the clients are networked with one another, and instead of being known as clients, the biometric devices become known as peers. For an example of a peer to peer network, see Figure 3.4.

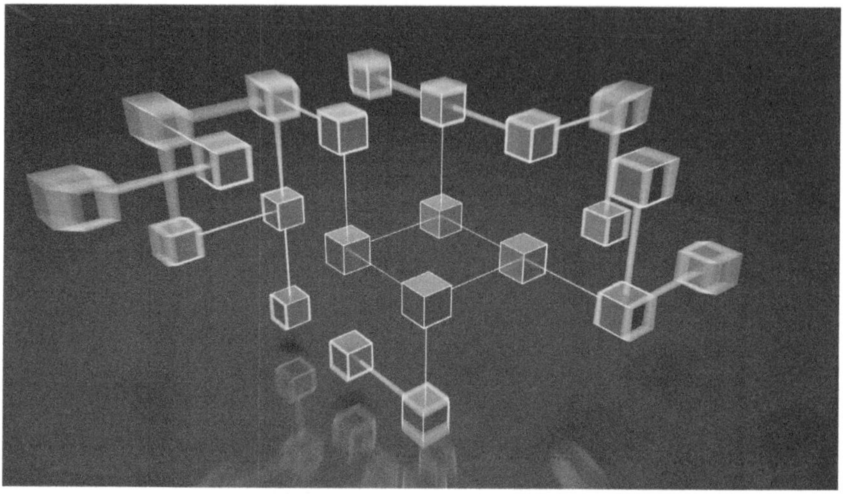

*Figure 3.4* This is an image of a peer to peer network. (A group of cubes that are on a black surface photo – free network image on Unsplash.)

For example, if the processing of verification and identification transactions become too much for one peer to handle, then the workload can be transmitted across the other peers, since they are all networked together. There are two key advantages of the peer to peer biometric network. They are self scalability, and cost savings.

For example, with the former advantage, each peer adds on processing resource capabilities because the workload can be shared amongst all of the peers, as opposed to just a few main hosts. Also, since processing of the verification and identification applications are conducted with each peer, the processing time is much quicker as opposed to the client–server biometric network topology. This is so because, the biometric templates and other relevant information and data has to be transmitted back to the host in order for the transaction processing to take place.

With regards to the latter advantage, since the peers are also hosts to a certain degree, there is no need to acquire a dedicated, central host. Also a combination of a peer to peer biometric network and a client–server biometric is possible, such a network is known as a hybrid biometric network.

As it was discussed previously, packet switching is often used to help the data packets follow the most optimal network route in order to reach the destination biometric device or a central server, in the most efficient and quickest manner possible.

However, packet switching is not used in most typical biometric system network architectures, rather it is only used in the largest of biometric system applications. A example of this, once again, is that of the large AFIS databases, or the very large international airport environments. Or for that matter, if a very large multinational corporation has a massive biometric network system layout, packet switching will most likely be used as well.

## Routers

But, to help further optimize the flow of data packets (which contain the biometric templates) in a very large biometric system network topology, routers are often used to help break down the maze of network connections from biometric device to biometric device, and from biometric device to central server. In other words, a router can be used to help literally create a group of smaller networks amongst a huge, overall, biometric system network topology. These smaller networks are known as subnets. An example of a router can be seen in Figure 3.5.

Apart from data packet optimization, routers are used also to help the biometric system administrator fully map out and understand the logical flow of data packets between biometric devices and central servers. As a result, the flow of data packets will be much more smoother and efficient, and because of this, the transaction processing times of both verification and identification applications will be also just that much quicker, especially in critical, large-scale law enforcement biometric template searches.

*Figure 3.5* This is an image of a router. (A white laptop computer sitting on top of a table photo – free router image on Unsplash.)

A router can be specifically defined as "a physical device that joins multiple networks together" (Source: http://en.wikipedia.org/wiki/Network_packet). Essentially, the primary goal of the router in a large-scale biometric system network topology is to forward the data packets containing the biometric templates along to the right destination biometric device or central server.

This is done by collecting the information and data from the header of the incoming data packets which are going into the router. Obviously, the router by itself is not going to know the most optimal path for to send the data packets along to its ultimate destination, so thus, it makes use of what is known as a routing table.

In technical terms, the router can be defined as

> data table stored in a router that lists the routes to particular network destinations, and in some cases, metrics (the distances) associated with those routes. It is a data file in RAM that is used to store route information about directly connected and remote networks. The routing table contains network/next hop associations.
>
> *Source: Computer Networking: A Top Down Approach, Kurose, J.F. and Ross, K.W., 2008, Pearson Education, p. 324*

## Routing tables

Put in simpler terms, the routing table is merely a map inside of the router which tells it where to send the incoming data packets upon which appropriate outgoing network link, in order to ensure that the data packet reaches the right destination biometric device or central server. A routing table can also calculate the total distance which is required to send the data packet to the destination biometric device or central server. To help even further optimize this, the shortest routes are utilized as much as possible.

A router consists of the following components:

1. Input ports: This is where the incoming data packets coming into the biometric system network topology first enter into, and it is the incoming point which also helps send the data packets into the right direction into the switching fabric;
2. Switching fabric: This piece of hardware is located directly inside of the router, and this connects and acts as a bridge between the input ports and the output ports;
3. Routing processor: This is like the CPU and the RAM of the router, this contains and manages all of the routing tables, and helps to a small degree, the other various network management tasks;
4. Output ports: This port sends the data packets onto its final destination, whether it be the destination biometric device, or the central

server. Of course before the data packet is sent out from here, the most optimal path first is calculated by the routing table.

## Network traffic collisions

Although it does not happen very often, primarily because the biometric network system topology is a highly specialized network which serves only a few main objectives, from time to time, data packets can get backed up at the point of entry into the incoming port. This can literally cause a traffic jam of sorts, thus further exacerbating the verification and identification processing times.

How does such a situation get resolved? There are three primary methods:

1. Tail drop: In simple terms, with this methodology, if the router becomes too overloaded with incoming data packets trying to seek entry into the biometric network system topology, the excess of data packets are merely dropped from the incoming queue;
2. Random early detection: With this technique, statistical probabilities are calculated and assigned to each and every incoming data packet. This is used to help determine which data packets should get dropped;
3. Weighted random early detection: In this technique, instead of statistical probabilities being calculated, statistical weights are computed and assigned to each incoming data packet to help ascertain which ones should get dropped.

Also, since a router has to examine the legitimacy of the incoming data packets, in order to ascertain where it needs to send them along in the biometric system network topology, a router can also act as a firewall, and help protect the overall biometric system network. This is done when the router discards and permanently deletes any malicious data packets from entering into the incoming port of the router.

## REVIEW

The last chapter of this book reviewed the key concepts of biometric system architecture, the entire project management lifecycle as it relates to biometric deployment, as well as some of the major networking topologies which can be found in everyday medium to large sized biometric applications. But, when one thinks of an actual biometric system, one only thinks of it as a system which provides security to the place of business or organization.

However, a biometric system has its particular "needs" as well, in terms of both maintenance, upgrades, as well as fine tuning. In fact, in some ways, a biometric system can be likened to that of the cardio vascular system of the human body. It is the heart which is consistently providing fresh blood to our circulatory system, and also which recycles the "used" blood into "fresh" or "good" blood once again.

But, the heart also needs fuel itself, via the coronary arteries which provides it with fresh blood to the heart to keep it nourished and healthy. It is these very same coronary arteries which also need to be protected, really even secured, so that they do not develop occlusions themselves. If this were to happen, a diminished flow of blood result, thus starving the heart of the nutrients it needs in order to act as an effective pump.

In this very same regard also, the biometric system needs to be secure itself, so it is not prone to any types or kinds of security breaches. When these breaches do indeed happen, this is also analogous to the occlusions which form in the coronary arteries. The common notion is that since a biometric system itself is a security tool, it does not need any further layers of security in of itself.

But just as described, this is far from the truth. As mentioned previously, at times, throughout this book, biometrics is just another piece of technology which is also prone to its fair share of hacks and security attacks. But exactly, where can these security vulnerabilities within a biometric system?

Some of the most common attacks can take place to the actual hardware or the software applications themselves. But normally, these attacks can for the most part be prevented in the future with regular maintenance of the biometric system by installing the required firmware and software upgrades and patches.

Also, hackers have tried to spoof biometric systems by presenting such scenarios as a fake fingerprint or even a fake eye in front of the biometric sensor. But also, as discussed earlier in this book, a biometric system only processes the physiological and behavioral traits from living tissues.

But, another area which is also prone to hacking and security breaches is in the biometrics templates, also. As it was reviewed in Chapter 2 of this book, biometric templates are just mathematical representations of our own unique physiological and behavioral traits, which is used for both verification and identification applications.

If a biometric templates are hacked into, at least in theory, there is not too much that a hacker can do with it-at least not yet. But, there is always a potential that a hacker can cause real harm with a hijacked biometric template-especially given the rapid developments of technology which are in their hands. Also, given the societal impacts that biometrics technology possesses (this will be reviewed and examined in more detail in Chapter 6), if a biometric database were to be hacked into, and the templates were indeed stolen, dears amongst the end users would greatly escalate, and also, acceptance of the biometric system would greatly deteriorate in turn.

Chapter 4

# Biocryptography

## CRYPTOGRAPHY AND BIOCRYPTOGRAPHY

Therefore, it is very important to further protect the biometric templates – by adding an extra layer of security to them. This solution is primarily answered by the science of cryptography – especially that of biocryptography, which combines both the sciences and technologies of biometrics and cryptography. Therefore, the goal of this chapter is to review the science and principles of cryptography, and how it lends itself to the emerging field of biocryptography.

## INTRODUCTION TO CRYPTOGRAPHY

Cryptography is a science which dates all the way to the back to the times of Julius Caesar. In its simplest terms, the science of cryptography is merely the scrambling and the descrambling of text, or written messages between two individual parties.

These individual parties can also be referred to the sender and the receiver. It is the former which creates the text or the written message which needs to be sent, and in turn, it is the latter which receives the text or the written message, and then reads it and appropriately responds.

In normal, everyday communications, we always trust that the individual party who is receiving the text or written message will receive it accordingly, without any type or kind of problem. While this does mostly happen in our daily lives, given especially the high tech world we live in today, this sometimes does not occur.

And, when this does actually happen, we always assume that the worst has always occurred. But, what is the worse that could happen? Well, the text or the written message could be intercepted by a third party and maliciously used. Now once again, in normal everyday conversations, while we would normally trust the other party (the receiving party) from keeping the details of the conversation privileged, there is always a chance that a third

DOI: 10.4324/9781003562146-4

party could be covertly listening in, and use that very privileged information for purposes of personal gain or exploitation, such as that of identity theft.

We can also extend this example to electronic communications of all types and kinds. For example, when we hit the "send" button, what assurances do we have that the receiving party will get our message, or that it will not be intercepted by a third party? Obviously, we cannot really ensure any type or kind of safety, especially when it comes to electronic communications, like of that of e-mail which gets transmitted all over the worldwide networks and the Internet.

But, the only thing that can be guaranteed is that if any type or kind of message were to be captured by a third party, it would be rendered useless. But, how is this task actually accomplished? It is done by the scrambling and the descrambling of the text or the written message. Much more specifically, the text or the written message is scrambled by the sending party, and it remains scrambled while it is in transit, until the receiving party gets the text or the written message.

## MESSAGE SCRAMBLING AND DESCRAMBLING

Then at this point, the text or the written message must be unscrambled in order for it to make comprehensible sense for the receiving party. For example, a very simple example of this is that of "I LOVE YOU". The sending party would scramble this message by rearranging the letters as "UYO I VEOL". This message would then stay in this scrambled format while it is in transit, until it is received by the receiving party.

They would then descramble it, so it would read once again "I LOVE YOU". So, if this message were to have been captured by a third party, the content would be rendered useless and totally undecipherable to third party. This, in very simple terms, is the science of cryptography. It is basically the art of scrambling, and in turn, the descrambling of the text or the written message into a readable and comprehensible format, yet once again.

Specifically, cryptography can be defined as

> the practice and study of techniques for secure communication in the presence of third parties (called adversaries). More generally, it is about constructing and analyzing that overcome the influence of adversaries and which are related to the various aspects of data confidentiality, data integrity, authentication, and repudiation.
>
> *Source: Computer Networking: A Top Down Approach, Kurose, J.F. and Ross, K.W., Pearson Education Group (2008, p. 683)*

As one can also discern form this definition that modern day intersects with other branches of study, especially that of mathematics, and computer sciences. And today, it is now even finding an intersection point with that of biometrics, into the new field known as biocryptography. This will be examined obviously in much more detail, later in this chapter, after some of the more fundamental concepts have been covered.

## ENCRYPTION AND DECRYPTION

In terms of cryptography, the terms of scrambling and descrambling have much more specific terms associated with them. Respectively, scrambling and descrambling are also known as "encryption" and "decryption". So for instance, the written message of "I LOVE YOU", when it is scrambled by the sending party becomes what is known as the "encrypted message", meaning that the written message has been disguised in such a manner that it would be totally meaningless, or in the terms of cryptography, it would be what is known as "undecipherable".

Also, encryption can also be further defined and described as "conversion of information from a readable state to apparent nonsense" (Source: *Computer Networking: A Top Down Approach*, Kurose, J.F. and Ross, K.W., Pearson Education Group, 2008, p. 683). Now, when the receiving party receives this encrypted written message, it must be descrambled into an understandable and comprehensible state of context. This process of descrambling is also known as "decryption".

So, rather than saying that cryptography is the science of scrambling and descrambling, it can now be referred to as the science of encryption and decryption. There also specific terms which are used for the encrypted message as well as the decrypted message. For example, the decrypted message, when it is returned back into its plain or original state of context which is comprehensible and decipherable is also known as the "cleartext" or the "plaintext".

### Ciphertexts

And, when the decrypted message is once again encrypted into a state of context which is totally incomprehensible and undecipherable, this is known as the "ciphertext". So, to illustrate all of this, with the previous example, when the sending party creates the written message of "I LOVE YOU", this is the plaintext or the cleartext.

Once this message is encrypted into the format of "UYO I VEOL", and while it is in transit, it becomes known as the ciphertext. Then, once the receiving party gets this ciphertext and then decrypts it into a comprehensible and understandable form of "I LOVE YOU", this message then becomes the plaintext or the cleartext, yet once again.

At this point, the question often gets asked is how does the sending party actually encrypt the text or the written message and how does the receiving party then actually decrypt the ciphertext (which is once again, the text or written message which is encrypted)?

Well, in its simplest form, the text or the written message is encrypted via a special mathematical formula. This formula is specifically known as the "encryption algorithm". Because the ciphertext is now encrypted by this special mathematical algorithm, it would be rendered useless to a third party with malicious intent, because of its totally garbled nature.

As the receiving party receives this ciphertext, it remains in its garbled format, until is it is descrambled. To do this, a "key" is used, which is only known by the sending party and the receiving party. In terms of cryptography, this key is also known as the "cipher", and it is usually a short string of characters, which is needed to break the ciphertext.

As it will be examined later in this chapter, interestingly enough, the encryption algorithm is actually publicly known and is available for everyone to use. Therefore, the key, or the ciphertext, must remain a secret between the sending party and the receiving party.

In order to send the ciphertext between the sending party and the receiving party, and as well to share the keys that is needed to encrypt and decrypt the ciphertext, specific cryptographic systems are needed. Today, there are two such types of cryptographic systems which exist. They are known as symmetric key systems and asymmetric key systems.

## SYMMETRIC KEY SYSTEMS AND ASYMMETRIC KEY SYSTEMS

The primary difference between these two types of cryptographic systems is that the former uses only one key for encryption and decryption, which is known as the private key of the ciphertext. With the latter, two types of keys are utilized for encryption and decryption of the ciphertext, and these are known as the public key and the private key.

We look at both of these cryptographic systems, first starting with symmetric key systems. One of the simplest methodologies in symmetric key systems is that of the Caesar cipher, which can be attributed to as far back as Julius Caesar (thus its name).

## THE CAESAR METHODOLOGY

With the Caesar methodology, each letter of the text or the written message is substituted with another letter of the alphabet which is sequenced by so many spaces, or letters later in the alphabet. To make things simpler to understand, we can denote this specific sequencing as "k". So, four letters out into the alphabet would be represented mathematically as "k=4".

Let us go back to our example once again of "I LOVE YOU". If k=4, which represents the letter sequencing, the letter that would replace "I" in this plaintext message would be the letter "E". So, continuing in this fashion, the ciphertext would be translated as "E YOUA EIM". With the Caesar cipher, a technique known as "wrap around" is possible. Meaning, once the last letter of the ciphertext is reached in the ciphertext is reached in the alphabet (which would be the letter "Z"), it wraps around immediately to the start of the alphabet, with the letter "A".

So, if the wrap around technique were needed to finish the encryption of this plaintext message, with k=4, the letter "A" in the plaintext message would become the letter "D", and so on. The value of the "k" serves as the key in the Caesar cipher, since, it specifies how the plaintext should be encrypted over into the ciphertext.

With the Caesar cipher, some 25 different combinations, or key values can be used. An improvement over the Caesar cipher came with a newer technique with what is known as the "monoalphabetic cipher". What distinguishes this from the Caesar is that, although one letter of the alphabet can still be replaced with another, there is no exact mathematical sequencing which is required.

Rather, the letters in the plaintext can be substituted at random in order to create the ciphertext. So once again, for example, the plaintext message of "I LOVE YOU" can be written at will and at random as "UYO VOLI E". With the monoalphabetic cipher, there are more pairings of letters possible. For example, there are $10^{26}$ possibilities of letter pairings versus only the 25 letter pairings available with the Caesar cipher.

Thus, if a hacker were to attempt a brute force attack on a monoalphabetic cipher (which is just the sheer guessing of the ciphertext for any type or kind of pattern in order to decipher the plaintext), it would obviously take a much longer time to crack versus the Caesar cipher.

## Types of cryptographic attacks

However, with both of these of types of cryptographic methods just described, there are three types of attacks they are vulnerable to, and are detailed as follows:

1. Ciphertext-only attack: With this type of attack, only the ciphertext is known to the attacker. But, if this particular individual is well trained in statistics, then he or she can use various statistical techniques, to break the ciphertext back into the plaintext;
2. Known-plaintext attack: This occurs when the hacker knows some aspect of either the letter pairings; thus, they can consequently crack the ciphertext back into the plaintext;
3. Chosen-plaintext attack: With this type of attack, the hacker can intercept the natural plaintext message which is being transmitted across the network medium, and from this, reverse engineer it back

into its ciphertext form, in an attempt to figure out the specific encryption scheme.

## Polyalphabetic encryption

Over time, improvements were made to both the Caesar cipher and the monoalphabetic cipher. The next step up from these two techniques was another technique known as the "polyalphabetic encryption". With this, multiple types of Caesar ciphers are used, but these ciphers are used in a specific sequence, which repeats once again when the overall cipher has reached its logical end the first time in order to finish the completion of the encryption of the plaintext message.

This means that the wrap around technique is also prevalent in this type of scenario. Let us illustrate this example once again with "I LOVE YOU". Building upon the example used previously, suppose that two types of Caesar ciphers are being utilized, such as where k=1, and k=2 ("k" once again denotes the actual Caesar cipher, or the sequential spacing of the number of letters later in the alphabet).

The following chart demonstrates this, in order to make it clearer:

| | |
|---|---|
| Plaintext: | A B C D E F G H I J K L M N O P Q R S T U V W X Y Z |
| 1st Caesar cipher, where k=1 | B C D E F G H I J K L M N O P Q R S T U V W X Y Z A |
| 2nd Caesar cipher, where k=2 | C D E G H I J K L M N O P Q R S T U V W X Y Z A B C |

The overall cipher algorithm utilized is C1 (k=1), C2 (k=2) where C denotes the Caesar key. So, with the example of using "I LOVE YOU", using the polyalphabetic algorithm, it would be encrypted as "J ORYF BPX". To understand this further, the first letter in the plaintext is C1, so I is represented as J, the second letter of the plaintext is C2, so C is represented as O, and so on.

The logical end of the cipher algorithm is C2, so once again, it reaches the logical end of its first iteration, it then wraps around once again as C1 (k= 1), C2 (k=2) the second time around, then the third time around, until the plaintext message has been fully encrypted.

So, in our illustration of "I LOVE YOU", there were a total of three iterations of C1 (k=1), C2 (k=2), in order to fully encrypt the plaintext. Over the years also, a more modern form of encryption was developed, known as the "block cipher". With this method, the plaintext message is put together in one long stream, and from this, it is then broken up into blocks of text of equal value.

## Block ciphers

Then from here, using a method of transposition, the plaintext message is then encrypted into its scrambled format. Let us illustrate this again with our example used before, but this time, let us assume a block of three characters, mathematically represented as 3 bits, or where k=3.

| | |
|---|---|
| Plaintext: | I LOVE YOU |
| Plaintext Block: | ILO VEY OUX |
| Ciphertext Block: | OLI YEV XUO |
| Ciphertext: | OLIYEVXUO |

Note that an extra character as added at the end, which is the letter "X". This was added so that a complete plaintext block can be formed. As a rule of thumb, if the total number of characters in the plaintext is not divisible by the block size permutation (in this instance, where k=3), then it can be safely assumed that extra characters will be needed to the plaintext in order for the last block of plaintext to be considered as complete. This is known as "padding". It should be noted that the most widely used block is where k=8 bits long.

As we can see, even with the simple example from up above, block ciphers are a very powerful tool for symmetric key cryptographic systems. After all, it goes through a set number of iterations of scrambling, in order to come up with a rather well protected ciphertext. But despite these strong advantages of block ciphers, it does suffer from one kind of inherent weakness.

And if, it is discovered by a hacker, it can cause rather detrimental damage, with irrevocable results. This vulnerability is that the two blocks of data can contain the exact same data. Let us examine this with our previous example once again. As it was illustrated, the ciphertext block was formulated as "OLI YEV XUO". But of course, depending upon the actual written context of the plaintext, it is possible that the ciphertext block can contain two or more exact blocks of the same data.

## Initialization vectors

So for example, it would look like this: "OLI OLI YEV". To alleviate this weakness, a system of Initialization Vectors (also known as IVs) is used. Although it sounds complex, simply put, it is the technique of creating some further scrambling, or randomness within the ciphertext block itself. But however, it should be noted that it is not the IV itself which does not further the ciphertext blocks.

## CIPHER BLOCK CHAINING

Rather, the IVs are part of a much larger process known as "Cipher Block Chaining", or CBC for short. Within this methodology, multiple loops of encryption are created in order to totally further scramble the ciphertext. Here is the how the process works:

1. The IV is created first;
2. Through a mathematical process known as XOR (which stands for eXclusive OR and is used quite frequently to determine if the bits of two strings of data match or not), the first created IV is XOR'd, with the first block of ciphertext data;
3. The first chunk of data which has been XOR'd is further broken down by another layer of encryption;
4. This process is then continued until all of the blocks of ciphertext have been XOR'd and enveloped with another layer of encryption.

Thus, this is how CBC gets its title. For instance, steps 1–4 as detailed up above create the first loop or chain, the second loop or chain is then next initiated, and so on, until the ciphertext has been fully analyzed and encrypted by this methodology.

## DISADVANTAGES OF SYMMETRIC KEY CRYPTOGRAPHY

Now that we have reviewed some of the basic principles of symmetric key cryptography, although it can be a robust system to use, it does suffer from three major vulnerabilities, which are as follows:

1. Key distribution;
2. Key storage and recovery;
3. Open systems.

With regards to the first one, key distribution, symmetric cryptography requires the sharing of secret keys between the two parties (sending and receiving), which requires the implicit trust that this key will not be shared with any other outside third party. The only way that any type of secrecy can be achieved in this regard would be to establish a secure channel.

While this works very well in theory, in practicality, it is not a feasible solution. For instance, the typical place of business or organization would not be able to afford implementing and deploying such a secure channel, except for the very large corporations and government entities. Thus, the only other solution available in this circumstance would be the use of a so-called designated controller.

This third party would have to be very much trusted by both the sending and the receiving parties. But, this methodology of trust to create a secure channel can prove to be a very cumbersome task. For example, imagine a place of business. Suppose also that the CEO decides to share the keys of the business with the employees who need access to it at irregular hours.

Rather than trusting the employees explicitly, the CEO could decide to utilize a manager to whom the employees must give the key when they are done with their job duties, and from there, this same manager would then give this key to the next employee who needed access.

Already, one can see that this is a very tedious and time consuming process, and to compound to this problem even more is that the designated controller, in this case the manager, cannot be trusted either, because in between the distribution of keys to the employees, this manager could very well give these same keys to a malicious third party. As a result, when this is applied to the world of cryptography, this method does not at all, by any means, guarantee the secrecy of the key which is needed to encrypt and decrypt the plaintext message.

In terms of key storage and capacity, let us take the example of a very large place of business or organization, such as that of a multinational corporation. The problem of using the principles of symmetric cryptography becomes quite simple.

First, since there will be many more lines of communication between the sending and the receiving parties, the need to implement that many more controllers (and perhaps more) becomes totally unrealistic as well as infeasible. Thus, the distribution of the keys can become a virtual nightmare.

Second, all of the private keys associated in symmetric cryptography have to be securely stored somewhere, primarily in a database which resides in a central server. As it is well known, primary and central servers are often very much prone to worms, viruses, as well as other types and kinds of malicious software. Compounding this problem even more is the fact that if there are many private keys stored onto this central server, the greater the chances of the central server being hacked into.

A way that these private keys can be stolen is if a piece of malicious code is injected into the intranet of the corporate network, which in turn reaches the database. This malicious code then actually very covertly hijack and send back to the hacker these private keys.

Third, when companies and organizations get large, the chances that employees will require remote access to the corporate intranet and network resources become even greater. As a result, the private keys which are used to communicate between the sending and receiving parties can also be hijacked very quickly and easily, by a hacker who has enough experience and knowledge into what they are doing.

Finally, with an open system, private or symmetric cryptography works best only when it is used in a very closed or "sterile" environment, where there are at best only just a few (or even just a handful) of sending and receiving parties. But, this is not the case with "open" or public environments, such as our example of the very large corporation. In these situations, there is simply no way to confirm the authenticity or the integrity of the private keys and their respective ciphertext messages.

So, as one can see, private keys and symmetric cryptography simply are inflexible, too costly, and do not scale well for most types and kinds of environments. For example, "solutions that are based on private-key cryptography are not sufficient to deal with the problem of secure communications in open systems where parties cannot physically meet, or where parties have transient interactions". (Source: *Computer Networking: A Top Down Approach*, Kurose, J.F. and Ross, K.W., Pearson Education Group, 2008, p. 687).

Although there will never be a perfect 100% solution which will correct the flaws of symmetric cryptography, there is a partial solution known as "Key Distribution Centers", which is reviewed next.

## THE KEY DISTRIBUTION CENTER

The key distribution center, also referred to as the KDC for short, is primarily once again a central server which is dedicated solely to the KDC network configuration. It merely consists of a database of all of the end users at the place of business or organization, and their respective passwords, as well other trusted servers and computers along the network.

It should be noted that these passwords are also encrypted. Now, if one end user wishes to communicate with another end user on a different computer system, the sending party enters their password into the KDC, using a specialized software called "Kerberos". When the password is received by the KDC, the Kerberos then uses a special mathematical algorithm which adds the receiving party's information and converts it over to a cryptographic key.

Once this encrypted key has been established, the KDC then sets up and establishes other keys for the encryption of the communication session between the sending and the receiving party. These other keys are also referred to as the "tickets". These tickets have a time expiration associated with them, so the ticket will actually expire at a predetermined point in time, in order to prevent unauthorized use, and it would also be rendered useless, if it is stolen, hijacked, or intercepted by a third party.

Although the KDC system just described does provide a partial solution to the shortcomings of symmetric key cryptography, the KDC also by nature has some major security flaws, such as:

1. Because the KDC contains all of the master keys and the access rules needed for encrypted communication, the server which contains the

KDC system must be both logically and physically protected all the time. If an attack is successful on the KDC, the entire communications channel within the place of business or organization will completely break down. Also, personnel whim have access to the KDC can easily decrypt the ciphertext messages between all of the sending and receiving parties.
2. The KDC process presents a single point of failure for the place of business or organization. If the server containing the KDC crashes, then all types and kinds of secure communications becomes impossible to have, at least on a temporary basis. Also, since all of the end users will be hitting the KDC at peak times, the processing demands placed onto the KDC can be very great, thus heightening the chances that very slow communications between the sending and the receiving parties, or even a breakdown of the communications system can also happen.

## Mathematical algorithms with symmetric cryptography

There are a number of key mathematical algorithms which are associated with symmetric cryptography, and they can be described as follows:

1. The Needham-Schroeder algorithm: This algorithm was specifically designed for KDC systems, in order to deal with sending and receiving parties from within the place of business or organization, who appear to be offline. For example, if the sending party sends a ciphertext message to the receiving party, and after sending the message they go offline, the KDC system could just literally "hang", and maintain an open session indefinitely, until the sending party comes back online again. With this particular algorithm, this problem is averted by immediately terminating the communication session once either party goes offline.
2. The Digital Encryption Standard (DES) algorithm: This mathematical algorithm was developed in 1975, and by 1981, it became the de facto algorithm for symmetric cryptography systems. This is a powerful algorithm, as it puts the ciphertext through 16 iterations in order to ensure full encryption.
3. The Triple Digit Encryption Standard (3DES) algorithm: This mathematical algorithm was developed as an upgrade to the previous DES algorithm just described. The primary difference between the two of them is that 3DES puts the ciphertext through three times as many more iterations than the DES algorithm.
4. The International Data Encryption Algorithm (IDEA): This is a newer mathematical algorithm than 3DES and is constantly shifting the letters of the ciphertext message around constantly,

until is decrypted by the receiving party. It is three times faster than any of the other DES algorithms just reviewed, and as a result, it does not consume as much processor power as the DES algorithms do.

5. The Advanced Encryption Standard (AES) algorithm: This is the latest symmetric cryptography algorithm and was developed in 2000, primarily designed for use by the federal government.

## THE HASHING FUNCTION

Finally, in symmetric cryptography, it should be noted that all of the ciphertext messages come with what is known as a "hash". It is a one way mathematical function, meaning, it can be encrypted, but it cannot be decrypted. Its primary purpose is not to encrypt the ciphertext, rather its primary purpose is to prove that the message in the ciphertext has not changed in any way, shape or form. This is also referred to as "message integrity".

For example, if the sending party sends it message to the receiving party, the message (or the ciphertext) will have a hash function with it. The receiving party can then run a hash algorithm, and if the ciphertext message has remained intact, then the receiving party can be assured that the message they have received is indeed authentic and has not been compromised in any way. But, if the hash mathematical values are different, then it is quite possible that the message is not authentic and that it has been compromised.

In the last section of this chapter, we reviewed at length the principles of symmetric cryptography. With this methodology, only one key is used to encrypt and decrypt the ciphertext between the sending and the receiving parties. Now, in this section, we look at an entirely different methodology – called asymmetric key cryptography. With this type of methodology, not just one key is used, but rather, two keys are used.

## ASYMMETRIC KEY CRYPTOGRAPHY

These keys are called the public and the private key and are also used to encrypt and decrypt the ciphertext which is sent between the sending and the receiving parties as they communicate with another. In the most simplistic terms, asymmetric cryptography can be likened to that of a safety box at a local bank. In this example, normally, there are two set of keys which are used.

One key is the one which the back gives to you. This can be referred to as the public key, because it is used over and over again by past renters of this particular safety deposit box, and for other, future renters as well. The second key is the private key which the bank keeps in their possession at all times and only the bank personnel know where it is kept.

The world of asymmetric cryptography is just like this example, but of course, it is much more complex than this in practice. To start off with, typically, in asymmetric cryptography, it is the receiving party which is primarily responsible for generating both the public and the private key. In this situation, let us refer to the public key as "pk", and the private key as "sk".

So, to represent both of these keys together, it would be mathematically represented as (pk, sk). It is then the sending party which uses the public key (pk) to encrypt the message they wish to send to the receiving party, which then uses the private key (sk) which they have privately and personally formulated to decrypt the encrypted ciphertext from the sending party.

Remember, one of the primary goals of asymmetric cryptography is to avoid the need for both the sending and the receiving parties from having to meet literally face to face in order to decide on how to protect (or encrypt) their communications with another. So, at this point, the question then arises is how does the sending party know about the public key (pk) generated by the receiving party so that the two can communicate with each other?

## Keys and public private keys

There are two distinct ways in which this can be accomplished: (1) The receiving party can deliberately and purposefully notify the sending party of the public key (pk) in a public channel, so that communications can be initiated and then further established; and (2) the sending party and the receiving party do not know anything about each other in advance. In this case, the receiving party makes their public key known on a global basis, so that whoever wish to communicate with the receiving party can do so, as a result.

Now, this brings up a very important point: the public key is literally "public", meaning that anybody can use it, even all of the hackers in the world, if this is so the case. So, how does asymmetric cryptography remain secure? It remains solely on the privacy of the private key (sk) which is being utilized. In these cases, it is then up to the receiving party now to share the private key (sk) with any other party, no matter how much they are trusted.

If the privacy of the secret key (sk) is compromised in any way, then the security scheme of asymmetric cryptography is then totally compromised. In order to help ensure that the private keys remain private, asymmetric cryptography uses the power of prime numbers. The basic idea here is to create a very large prime number as a product of multiplying two very large prime numbers together.

Mathematically put, the basic premise is that it will take a hacker a very long time to figure out the two prime number multiples of a very large product which is several hundred integers long, and thus, give up in frustration. Even if a hacker were to spend the time to figure out one of these prime numbers, the hacker still has to figure out the other prime number and the chances that they will figure this out is almost nil.

As a result, only one portion of the (pk, sk) is figured out, and the asymmetric cryptography technique utilized by the sending and the receiving parties still remains intact and secure. In other words, the hacker cannot reverse engineer one key to get to the other key to break the ciphertext. It should also be noted than in asymmetric key cryptography, the same public key can be used by multiple, different sending parties to communicate with the single receiving party, thus forming a one to many, or 1:N mathematical relationship.

## THE DIFFERENCES BETWEEN ASYMMETRIC AND SYMMETRIC CRYPTOGRAPHY

Now, that we have provided a starting point into asymmetric cryptography, it is now important at this juncture to review some of the important distinctions and the differences between this and symmetric cryptography. First, with symmetric cryptography, the complete 100% secrecy of the key must be assured. Whereas, as it has been just described, asymmetric cryptography requires only half of the secrecy, namely that of the private key (sk).

Although this might seem like just a minor difference, the implications of this has great magnitudes. For example, with symmetric cryptography, both the sender and the receiver need to be able to communicate the secret key generated with each other first and the only way that this can happen is if both parties met face to face with each other, before the encrypted communication can take place between both parties.

And, to complicate matters even more, it is absolutely imperative that this private, or secret key is not shared with anybody else, or even intercepted by a third party. But whereas, once again in asymmetric cryptography, the public key can be shared virtually indiscriminately with each other, without the fear of compromising security.

Second, symmetric cryptography utilizes the same secret key for the encryption and decryption of the ciphertext, but with asymmetric cryptography, two different keys (namely the public and the private keys) are both used for the encryption and the decryption of the ciphertext.

In other words, in asymmetric cryptography, the roles of the sender and the receiver are not interchangeable with one another like symmetric cryptography. This means that with asymmetric cryptography, the communication is only one way. As discussed, because of this, multiple senders can send their ciphertext to just one receiver, but in symmetric cryptography, only one sending party can communicate with just one receiving party.

Also, asymmetric cryptography possesses two key advantages: (1) it allows for the sending party(s) and the receiving party to communicate with another, even if their lines of communication are being observed by a third party; and (2) because of the multiple key nature, the receiving party needs to keep only one private key to communicate with the multiple sending parties.

## The disadvantages of asymmetric cryptography

But despite all of this, asymmetric cryptography does possess one very, serious disadvantage: When compared to symmetric cryptography, it is two to three times much slower than symmetric cryptography. This is primarily so because of the multiple parties which are involved, and the multiple keys which are involved also.

Thus, this takes enormous processing power and is a serious drain to server power and system resources. So far in our review of asymmetric cryptography, we have assumed that the potential hacker is merely just eavesdropping in on the ciphertext communications between the sending and the receiving parties. But, if the potential hacker has a strong criminal intent, they can quite easily listen in on the communications on an active basis and cause great harm in the end.

There are two specific cases in which this can happen. First is the situation if the hacker replaces a public key of this own (mathematically represented as pk'), while the ciphertext is in transit between the sending and the receiving parties, and the receiving party decrypts that ciphertext with that malicious public key (pk'). But keep in mind, this scenario assumes that the hacker has some substantial information about the private key (sk).

The second situation arises when the hacker can change the mathematical value of the public key or change it while it is in transmission between the sending and the receiving parties. The exact techniques for protecting the public key in asymmetric cryptography will be discussed later in this chapter.

## The mathematical algorithms of asymmetric cryptography

There are number of key mathematical algorithms which serve as the crux for asymmetric cryptography, and of course, use widely differing mathematical algorithms than the ones used with symmetric cryptography. The algorithms used in asymmetric cryptography are as follows:

1. The RSA algorithm;
2. The Diffie–Hellman algorithm;
3. The Elliptical Wave Theory algorithm.

In terms of the RSA algorithm, this is probably the most famous and widely used asymmetric cryptography algorithm. In fact, this very algorithm will serve as the foundation for biocryptography later in this chapter. The RSA algorithm originates from the RSA data security corporation and is named after the inventors who created it, which are: Ron Rivest, Adi Shamir, and Leonard Adelman.

As it was just reviewed previously, the RSA algorithm uses the power of prime numbers to create both the public key and the private key. But using such large keys to encrypt such large amounts of data is totally infeasible from the standpoint of the processing power and central server resources. Instead ironically, the encryption is done symmetric algorithms (such as the ones reviewed previously), then the private key gets further encrypted by the receiving party's public key.

Once the receiving party obtains their ciphertext from the sending party, then the private key generated by the symmetric cryptography algorithm is decrypted, then the public key which was generated by asymmetric cryptography can then be subsequently used to decrypt the rest of the ciphertext.

In terms of the Diffie–Hellman asymmetric algorithm, it is named after its inventors as well, who are Whit Diffie and Martin Hellman. It is also known as the DH algorithm for short as well. But interestingly enough, this algorithm is not used for the encryption of the ciphertext, rather the main concern of it is to address the problem for finding a solution of the issue of sending a key over a secure channel.

Here is a summary of how it works, on a very simple level:

1. The receiving party as usual has the public key and the private key that they have generated, but this time, they both are created by the DH algorithm;
2. The sending party receives the public generated by the receiving party and uses this DH algorithm to generate another set of public keys and private keys but on a temporary basis;
3. The sending party now takes this newly created temporary private key and the public key sent by the receiving party to generate a random, secret number – this becomes known as the "session key";
4. The sending party uses this newly established session key to encrypt the ciphertext message, and sends this forward to the receiving party, with the public key that they have temporarily generated;
5. When the receiving party finally receives the ciphertext from the sending party, the session key can now be derived mathematically;
6. Once the above step has been completed, the receiving party can now decrypt the rest of the ciphertext.

Finally, with Elliptical Wave Theory, it is a much newer type of asymmetric mathematical algorithm. It can be used to encrypt very large amounts of data, and its main advantage is that it is very quick and does not require a lot of server overhead or processing time. As its name implies, Elliptical Wave Theory first starts with a parabolic curve drawn on a normal x, y coordinate Cartesian plane.

After the first series of X and Y coordinates are plotted, various lines are then drawn through the image of the curve, and this process continues until many more curves are created and their corresponding, intersecting lines are also created.

Once this process has been completed, the plotted X and Y coordinates of each of the intersected lines and parabolic curves are then extracted. Once this extraction has been completed, then all of the hundreds and hundreds of X and Y coordinates are then added together in order to create the public and the private key. But, the trick to decrypting a ciphertext message encrypted by Elliptical Wave Theory is that the receiving party has to know the shape of the original elliptical curve, and all of the X and Y coordinates of the lines where they intersect with the various curves, and the actual starting point at which the addition of the X and Y coordinates first started.

## THE PUBLIC KEY INFRASTRUCTURE

Since the public key has become so important in the encryption and the decryption of the ciphertext messages between the sending and receiving parties and given the nature of the of its public role in the overall communication process, great pains, and extensive research have been taken to create an infrastructure which would make the process of creating and sending of the public keys as well as the private keys much more secure and robust.

In fact, this infrastructure is a very sophisticated form of asymmetric cryptography, and it is known as the "Public Key Infrastructure", or "PKI" for short. The basic premise of PKI is to help create, organize, store, distribute, and maintain the public keys. But, in this infrastructure, both of the private and public keys are referred to as "Digital Signatures", and they are not created by the sending and receiving parties, rather they created by a separate entity known as the "Certificate Authority".

This entity is usually an outside third party which hosts the technological infrastructure which is needed to initiate, create, and distribute the digital certificates. In a very macro view, the PKI consists of the following components:

1. The certificate authority, also known as the CA: This is the outside third party who issues the digital certificates;

2. The digital certificate: As mentioned, this consists of both the private key and the public key, which are issued by the CA. This is also the entity that the end user would go to in case he or she needed to have a digital certificate verified. These digital certificates are typically kept in the local computer of the employee or even the central server at the place of business or organization;
3. The LDAP or X.500 directories: These are the databases which collect and distribute the digital certificates from the CA;
4. The registration authority, also known as the RA: If the place of business or organization is very large (such as a multinational corporation), this entity then usually handles and processes the requests for the required digital certificates and then transmits those requests to the CA to process and create the required digital certificates.

In terms of the CA, in extremely simple terms, it can be viewed as the main governing body, or even the king of the PKI. In order to start using the PKI to communicate with others, it is the Certificate Authority which issues the digital certificates, which consists of both the public and private keys.

## THE DIGITAL CERTIFICATES

Each digital certificate which is generated by the Certificate Authority consists of the following technical specifications:

1. The digital certificate version number: Typically, it is either version number 1, 2, or 3;
2. The serial number: This is the unique ID number which separates and distinguishes a particular digital certificate from all of the others (this can be likened to each digital certificate having its own Social Security Number);
3. The signature algorithm identifier: This contains the information and data about the mathematical algorithm used by the Certificate Authority to issue the particular digital certificate;
4. The issuer name: This is the actual name of the Certificate Authority which is issuing the digital certificate to the place of business or organization;
5. The validity period: This contains both the activation and deactivation dates of the digital certificates, in other words, this is the lifetime of the digital certificate as determined by the Certificate Authority;
6. The public key: This is created by the Certificate Authority;
7. The subject distinguished name: This is the name which specifies the digital certificate owner;

8. The subject alternate name email: This specifies the digital certificate's owner email address (this is where the actual digital certificates go to);
9. The subject name URL: This is the web address of the place of business or organization to whom the digital certificates are issued to.

## HOW PUBLIC KEY INFRASTRUCTURE WORKS

This is how the first part of the PKI works:

1. The request for the digital certificate is sent to the appropriate Certificate Authority;
2. After this request has been processed, the digital certificate is issued to the person who is requesting it;
3. The digital certificate then gets signed by confirming the actual identity of the person who is requesting that particular digital certificate;
4. The digital certificate can now be used to encrypt the plaintext into the ciphertext which is sent from the sending party to the receiving party.

The Registration Authority is also known as the RA, as eluded to before. The RA is merely a subset of the CA, or rather, it is not intended to replace or take over the role of the CA; instead, it is designed to help if it becomes overwhelmed with digital certificate request traffic.

However, the RA by itself does not grant any type or kind of digital certificates, nor does it confirm the identity of the person who is requesting the digital certificate. Rather, its role is to help process the requests until the processing queue at the CA becomes much more manageable.

The RA sends all of the digital certificate requests in one big batch, rather than one at a time. This is process is known as "chaining certificates". The RA is typically found in very large, multinational corporations, where each office location would have its own RA, and the CA would reside at the main, corporate headquarters.

Finally, all digital certificate requests are processed by the RA are also associated with a chain of custody trail, for security auditing purposes. The RA can be viewed as a support vehicle for the CA, in which a mathematical, hierarchal relationship exists.

### Public key infrastructure policies and rules

It should be noted at this point that in order for either the Certificate Authority or the Registration Authority to function properly, it is important to have a distinct set of rules policies in place. These surround the use of the issuance, storage, and the revocation of the expired digital certificate. While it is out of the scope of this book to get into the exact details of all these rules and policies, the following is just a sampling of some of the topics which need to be addressed:

1. Where and how the records and the audit logs of the Certificate Authority are to be kept, stored, and archived;
2. The administrative roles for the Certificate Authority;
3. Where and how the public keys and the private keys are to be kept, stored, and also backed up;
4. What is the length of time for which the public keys and the private keys will be stored;
5. If public or private key recovery will be allowed by the Certificate Authority;
6. The length of the time of the validity period for both the public keys and private keys;
7. What will be the technique in which the Certificate Authority can delegate the responsibilities to the Registration Authority
8. Will the digital certificates to be issued by the Certificate Authority be used for applications and resources;
9. If the digital certificates to be issued by the Certificate Authority be used for the sole purposes of just encryption of the ciphertext;
10. If there are any types or kinds of applications which should be refused to have digital certificates;
11. When a digital certificate is initially authorized by the Certificate Authority, if there will be a finite period of time when the digital certificate will be subject to revocation.

As one can see, based upon the establishment of the many rules and policies which need to be set into place, the actual deployment and establishment of a PKI can become quite complex, depending upon the size and the need of the particular business or organization.

## The LDAP protocol

In terms of the database structure for the digital certificates, this is most useful and effective when the Lightweight Directory Active Protocol (LDAP) servers are utilized. LDAP is merely an acronym which stands for "Lightweight Directory Access Protocol" and is simply a database protocol which is used for the updating and searching of the directories which run over the TCP/IP network protocol (this is the network protocol which is primarily used by the PKI Infrastructure).

It is the job of the LDAP server of the PKI to contain such information and data as it relates to the digital certificates, the public and the private key storage locations, as well as the matching public and private key labels.

The Certificate Authority uses a combination of the end user name and the matching tags to specifically locate the digital certificates on the LDAP server. From that point onwards, it is the LDAP server which then checks to see if the requested digital certificate is valid or not, and it if it is valid, it then retrieves from its database a digital certificate which can then be sent to the end user. Although all digital certificates which are issued have a finite lifespan when they are first issued, they can also be revoked for any reason at any time by the PKI Administrator.

In order to accomplish this very specific task, a Certificate Revocation List or a CRL is used. This list is composed of the digital certificate serial numbers which have been assigned by the Certificate Authority. But looking this type of information and data can be very taxing on the system resources and processes.

Therefore, in this regard, it is obviously much easier to reissue the digital certificates as they expire, rather than to revoke them and then having to reissue them again, and of course, which would mean that the PKI System Administrator would then have to update the Certificate Revocation List.

## The public cryptography standards

The PKI is also governed by a body known as the Public Key Cryptography Standards, also known as the PKCS. The first of these standards has been previously described, and they are the RSA Encryption Standard, the Diffie–Hellman Key Agreement Standard and the Elliptical Wave Theory. The other sets of standards which define a PKI are as follows:

1. The password-based cryptography standard: This describes how to encrypt a private key with a secret key which is derived from a password;
2. The extended certificate syntax standard: This is merely a set of attributes attached onto a digital certificate which has been assigned by the Certificate Authority;
3. The cryptographic message syntax standard: This standard specifically outlines how to put the digital signatures into digital certificate envelopes, and from there, put that into another digital envelope;
4. The private key information syntax standard: This standard directly specifies what kind of information and data should be included into a private key, and how that specific key should be formatted;
5. The selected attribute types: This is a detailed list which at length describes the certain encryption attribute types for the last three standards;

6. The certification request syntax standard: This provides the details for the syntax for the digital certificates. Essentially, this standard simply sets forth the parameters which are needed for the Certificate Authority to understand the digital certificate request;
7. The cryptographic token interface standard: This is an Application Programming Interface (API) for specifying and handling the cryptographic functions as it relates to the smartcards;
8. The personal information exchange standard: This standard specifies exactly how an end user's private keys should be transported across the network medium;
9. The cryptographic token information format standard: This standard specifies how the applications at a place of business or organization should interface with smart cards.

In the world of PKI, it should be remembered that the public keys and the private keys (also known as the digital certificates) are created instantaneously and all of the time. In fact, public keys and private keys are everywhere in a PKI, even when one establishes a Secure Shell (SSH) connection over the Internet with their particular brand of web browser (this typically uses 128-Bit Encryption).

In fact, there are even public keys and private keys in the PKI which are only used once, terminated, and are literally discarded away. These types of public keys and private keys are known more commonly as "session keys". Public keys and private keys are nothing more than computer files.

## Parameters of public keys and private keys

But before the actual public key or the private key can go out, it needs to have certain parameters which are specified to it, before it can be created, and these are as follows:

1. The type of mathematical algorithm which should be used (as described previously);
2. How many bits of data the public keys and the private keys should be composed of;
3. The expiration date of both the public keys and the private keys.

Also, in order to keep the hackers at bay, it is equally important that not all of the public keys and the private keys are used all the time in the communication process between the sending and the receiving parties. It is also important to keep the public keys and the private keys fresh, or in other words, it is important to introduce randomness into the PKI.

Such randomness is known as "Entropy", and this entropy is created by what is known as "Random Number Generators", and "Psuedo-Random

Number Generators". Also, in a PKI, there are different classes of both public keys and private keys. Here is a listing of just some of these classes of keys:

1. Signing keys: These are the keys to create the digital signatures;
2. Authentication keys: These are the keys which are created to authenticate computers, servers, and the receiving parties and the sending parties with one another;
3. Data encryption keys: These are the keys which are used to encrypt the files;
4. Session keys: These are the types of keys which are used to help secure a channel across an entire network for only a very short period of time;
5. Key encryption keys: These types of keys literally wrap the ciphertext to provide further protection between the sending and the receiving parties;
6. Reof key: This is the master which is used for signing all of the other public keys and private keys which originate specifically from the Certificate Authority;

## How many servers?

In the PKI, especially in the small to medium-sized businesses, very often only one server is utilized in order to distribute both the public keys and the private keys to the employees within the place of business or organization.

But, the primary disadvantage with this is these types of key servers can literally become the single point of failure, if the server breaks down, or worst yet, it is even hacked into.

In order to help alleviate this problem, these types of business can have multiple, redundant servers, but this too can become a huge expense, especially for the small to medium sized business. Thus, as a result, the best option for the small- to medium-sized businesses as well as even the large multinational corporations is to outsource to t third party the entire PKI, such with such organizations as VeriSign.

This approach can be considered as a hosted one, and can greatly save on IT related expenses, and security is at the forefront of these hosting parties. Another key issue which is related to that of public key and private key distribution in a PKI is that of the setting the security policies for both the public keys and the private keys, and which mechanisms are used for their storage, as well as even furthering the securing of the public keys and the private keys themselves (this is a topic for what is known "Virtual Private Networks", or "VPNs", and this will be discussed in the next major section of this chapter).

## Security policies

The exact mechanisms as to how to exactly establish a specific security policy is not the realm of this book, but the security policy should cover at minimum, the following key issues, as it relates to public key and private key generation and distribution:

1. Who are the individuals which are authorized to access the key server (assuming that the place of business or organization has one);
2. Who within the place of business or organization is even allowed to use the public keys and the private keys at all;
3. What types and kinds of ciphertexts and content messages and even corporate data can use the encryption methods which are provided by the public and the private keys;
4. If the public key and private key generation and distribution processes are outsourced to a third party, who at the place of business or organization has the authority to issue public keys and private keys after they have been generated;
5. What are the specific requirements for the employees within the place of business or organization to obtain both public keys and private keys.

## Securing the public keys and the private keys

Related to the issue of both public key and private key security is that of how to secure the keys themselves. Broadly speaking, there are two ways in which this can be handled:

1. Key Escrow;
2. Key Recovery.

The former refers to the storage of the public keys and the private keys at a safe location, and the latter refers to breaking up the public keys and the private keys at the point of origination (which would be when the ciphertext is sent from the sending party), and putting them back together once again at the point of destination (which would be when the receiving party receives the ciphertext).

## Message digests and hashes

However, in today's business world, both of these methods are totally infeasible due to security concerns. Today, the preferred method by most Chief Information Officers (CIOs) is that is using message digests and hashes.

Both of these refer to the same concept and are often used synonymously with each other. Essentially, all it is a fixed length of literally data nonsense, such as:

> RTYHDHDHjjjdd8585858hd0909344jdjdjdjMNGDfsweqwecbthrdn *&^%gh$

This is to help ensure the integrity of the ciphertext while it is in transit across the network medium. In other words, this is proof positive for the receiving party who receives the ciphertext from the sending party that the ciphertext has remained intact while it has been in transit, and it has not been altered in any way shape or form.

The hash or the message digest can be viewed as a fingerprint of the ciphertext. The hash is actually created at the point of receiving party, and it is then calculated again via a mathematical algorithm utilized by the receiving party.

If the mathematical algorithm used by the receiving party generates the same type of garbled data message such as the one shown in the example, then receiving party can be 100% sure that the ciphertext they have received from the sending party is the original ciphertext at the point of origination and has remained intact.

## Security vulnerabilities of hashes

However, a major security vulnerability of using hashes is that they can even be altered while it is en route. In other words, a hacker can intercept the ciphertext and its associated hash, alter both, and create a brand new ciphertext and a brand new hash.

As a result, the receiving party is then fooled into believing that this new, altered ciphertext, and that the new altered hash is the original sent by the sending party, while the hacker keeps the actual ciphertext and hash which was generated the first time around.

To fix this major security vulnerability, the ciphertext is combined with a "secret key" at the point of origination first, then the hash is created. As a result, this hash will then contain specific information and data about the secret itself. As a result, the receiving party can even be further ensured that the ciphertext they have received is the 100% original sent by the sending party.

This is so because even if the ciphertext, the hash, and the associated secret key were to be intercepted, there is very little that a hacker can do to alter the ciphertext and its associated hash, because they have to have the information and data about the secret key, which is of course something they will never gain access to.

## Virtual private networks

As it was eluded to before, VPNs can be used as a much more sophisticated form of protecting the integrity of the ciphertext when it is sent from the sending party to the receiving party and vice versa. A VPN is essentially a dedicated network in of itself. It is highly specialized, with the main intention of highly securing the flow of communication between the sending and the receiving parties. An example of a VPN can be seen in Figure 4.1.

In its most simple form, the VPN takes the data packet in which the ciphertext resides (the ciphertext cannot just be sent by itself-it needs to have a vehicle to travel across the network medium, which is namely the data packet), and then further encrypts the data packet as needed.

This encryption is actually just another data packet. Meaning, the data packet which contains the ciphertext is further wrapped inside by another data packet, in order to provide that extra layer of protection. This process is technically known as "encapsulation".

Once this process has been accomplished, the VPN then establishes a dedicated network connection, or even a dedicated network channel, in which the encapsulated data packet can be sent in. Although this specific connection makes use of the public network infrastructure and related systems, this special type of network connection established and used by the

*Figure 4.1* This is an image of a virtual private network (VPN). (100+ CCTV camera pictures [HD] | Download free images and stock photos on Unsplash.)

VPN cannot be seen by others, as this connection cannot be picked up by network sniffers.

## IP tunneling

The establishment of this special type of network connection by the VPN is also known as "IP Tunneling". The following example best demonstrates how the scheme of IP tunneling actually works. First, assume that there is an employee with a laptop who wishes to connect to an internal server which is located deep within the company network.

Further assume that this laptop the employee possesses has a TCP/IP address of 1.2.3.4, and that the corporate server the employee wishes to connect to has a TCP/IP address of 192.168.1.1, and that it cannot be reached by any other public access mechanism.

In order for the individual's laptop to actually reach the internal corporate server, it must first penetrate a VPN firewall/server which has a public address of 10.11.12.13, as well as possessing a private TCP/IP address of 192.168.1.2. In order to fully establish the connection between the individual's laptop and the internal corporate server, a VPN Tunnel must first be established.

Here are the steps as to how this is accomplished:

1. The VPN client will connect to the VPN server/firewall through a specific network interface;
2. The VPN firewall/server will allocate a brand new TCP/IP address (such as 192.170.2.60), from the VPN's server/firewall subnet. This particular technique creates the virtual tunnel through which the encrypted data packet can reach then of the other VPN tunnel endpoint;
3. Now, when the individual's laptop (also known as the VPN client) wishes to communicate with the corporate server, it prepares its data packets to be sent to the TCP/IP address of 192.168.1.1, the data packets are then encrypted, and then further encapsulated with an outer VPN packet (such as IPSec packet), and relays these further, enhanced data packets to the VPN Firewall/Server with the public TCP/IP address of 10.11.12.13, over the public internet. This extra layer of protection is so well encrypted that even if a hacker were to intercept these particular data packets, there is nothing that they can do to reverse engineer the data packets in order to garner the information and data which resides in them;
4. In this stage, when the encapsulated data packets reaches the VPN Firewall/Server, it then extracts the inner data packets, decrypts them, and then further notes the destination TCP/IP address is 192.168.1.1, which is the actual corporate server, and from this points onwards, forwards these particular data packets to the corporate server.

This is the process of how this particular employee's laptop can be reach to, connect, have access, and communicate with the corporate server which resides deep in the heart of the corporate network. As one can see, the beauty of the VPN is that to an outsider, particularly a hacker, it looks and appears as if the communications of the data packets (such as in this example, between the employee's laptop and the corporate server) is transpiring across the same network segment, when in reality, the actual and real communications takes place across hidden, secret, and covert communications paths.

## Mobile virtual private networks

As the explosion of smartphones takes place in today's world, and especially in today's corporate setting, the use of what is known as mobile VPNs will also soon proliferate at an equally greater pace.

The traditional VPN model is a stationary one, meaning it can only be used at one type or kind of geographic location, such as the one depicted in the example up above. But however, in today's wireless world, stationary VPN's and the technology which is associated with them simply will not make it the quick and fastening pace of mobile technology.

Another perfect example of this is that of mobile devices used by law enforcement agencies. Today, law enforcement officers are equipped with various types and kinds of mobile devices which allow them to confirm the identity of the suspect of whom that they have apprehended. But rather than using the traditional ink and paper approach, biometrics is being used in order to provide for a quick determination and turnaround time for identification of the apprehended suspect.

In these particular mobile applications, the endpoint of the VPN is not fixed upon and located at a single TCP/IP address, but instead, the VPN connections, rather roams across the various networks such as data networks from cellular carriers or between multiple Wi-Fi access points.

In these cases, if the traditional VPN model were to be utilized, the virtual private connections would be keep continuously get broken, at each different location.

From this point onwards, it is then the Mobile VPN software which handles the necessary network authentication and also maintains the network tunneling sessions in a way which is totally transparent to the mobile application and the end user.

## Is the virtual private network worth it?

To the very large multinational corporations, and especially to the CFO's and the CIO's at these places of businesses and organizations, the ability to afford a high-end VPN solution comes to them at virtually no cost. Not literally speaking though, there is a true cost which is incurred by these businesses and organizations, but for these types of companies, purchasing such a high-end system is just merely a capital expenditure which can be easily written off as a mere deduction when it comes to report taxes in the next fiscal year.

But, for the small to medium sized businesses, obtaining a high-end VPN solution can be a very expensive proposition to undertake. For the CFO's and the CIO's of these much smaller places of business and organizations, obtaining a high level VPN solution can be an extraordinary drain on an already highly constrained and limited IT budget.

But, as the world is now turning more towards the use of biometrics-based VPN solutions, these small- to medium-sized businesses and organizations, at some point in time or another, will have to acquire a VPN solution just to keep at minimum, with the cat and mouse game between the end users, the vendors, and the hackers.

## Conducting a virtual private network cost benefit analysis

So, what do the CIO's and the CFO's have at their arsenal in order to help them to decide which VPN solution is affordable but will also be the best for their place of business or organization. One such technique is a very commonly used financial analysis methodology called a "Cost-Benefit Analysis", or CBA for short.

As it relates to the VPN, as the CFO or CIO for your business or organization, you first need to determine if you even have the funds needed to acquire a biometrics-based virtual private solution. The key financial questions to be asked and analyzed include the following items:

1. Determine your actual operating costs to run your current IT infrastructure;
2. Calculate the real and the tangible benefits that the biometrics-based VPN solution will bring to your business or organization as opposed to your existing IT security infrastructure;
3. Calculate the capital requirements as well as the operational costs of the new biometrics-based VPN solution;
4. Spread the financial costs of the estimated life of the new biometrics-based VPN solution, but it is most important to assign the capital expenditure costs to the specific years;
5. Ascertain and determine the cost differences between the operating costs of the existing security system in place, and the brand new biometrics-based VPN solution;
6. Based upon the financial numbers calculated above, now you can easily calculate the Return on Investment, or ROI, as well as the

Internal Rate of Return, or IRR, of the biometrics-based VPN solution, as well the as the Net Present Value. If both of the key financial statistics reveal a positive number, then the biometrics-based VPN is definitely worth at least considering to acquire.

Now, with the above key financial questions solidly answered, as the CFO and the CIO, you can now conduct the true Benefit Cost Analysis, and this should include the following components:

1. All of the costs which are associated to keep the biometrics-based VPN running and the associated upkeep costs;
2. All of the capital expenditures related to the lifetime of the biometrics-based VPN project;
3. The actual costs of hiring and keeping the employees to maintain the VPN system;
4. The calculations of the new revenues that the biometrics-based VPN system will bring to the place of business or organization;
5. Any other hidden costs it will take to implement the new VPN system;
6. Any other IT Budget cost savings which can be garnered by the new biometrics-based VPN system.

With all of this mind, as the CIO and the CFO, it is also important to keep in mind to stay from what is known as the "scope-creep" phenomenon. In other words, just implement the biometrics-based VPN system for the purposes it was intended to do, and not any more.

Otherwise, your operating costs will get greatly out of hand, In other words, the rules of IT project management must be strictly enforced. As the CIO, once you have discovered through the above processes detailed above, that a biometrics-based VPN solution is best for your place of business or organization, you must follow a very orderly approach in deploying the VPN system.

## Implementing a virtual private network

Simply throwing the system at the IT team and having them figure out how to deploy the actual system at your place of business or organization will simply not work. Rather, a very strong planned and iterative approach needs to take place. Such a controlled process involves the following steps:

1. Analyzing;
2. Pilot testing;
3. The limiting of the full blown biometrics-based VPN only in stages;

4. Conducting a complete VPN rollout once the stages as described up above have been completed, and everything is set in place.

It should be kept in mind that part of this iterative process includes conducting a very detailed systems analysis and requirements, which includes carefully reviewing and examining the following major requirements:

1. VPN security requirements;
2. VPN application program requirements;
3. Virtual private network user access requirements;
4. Virtual private network requirements;
5. VPN performance requirements.

In terms of the security requirements for your new biometrics-based VPN System, as the CIO, you will need to have an extremely firm grasp of the following:

1. An understanding of the types and kinds of information system requirements;
2. A classification of the information technology assets to actually determine what needs to be protected.

In terms of the Information Assets classification for the biometrics-based VPN, the key item to be determined is that of the tangible value of the assets-to three very important entities:

1. Your place of business or organization;
2. Your business associates (such as your suppliers);
3. Most importantly, your customers.

To fully understand and quantify the value of your information technology assets, as the CIO, there are three types of risks need to be ascertained as well, and they are as follows:

1. Integrity risk: The data packets which get sent through the VPN System have qualities of both as a commodity and a weapon, such as:
   a. Data loss;
   b. Data alteration;
   c. Data theft.
2. Confidentiality risk: The data packets which get transmitted back and forth between the VPN solution can contain very private and confidential data about the place of business or organization and are very prone to get lost, stolen, or hacked into and used for malicious purposes by a hacker;

3. Availability risks: This is a grave risk and can occur when the biometrics-based VPN totally becomes unavailable to the place of business or organization, and this is known as a Denial of Service Attack, when a hacker totally floods the VPN System and overloads it.

## The components of a virtual private network security policy

In implementing a biometrics-based VPN system, formulating and developing a very sound and "airtight" security policy becomes of prime importance. The components of a very strong VPN-based security policy should include some of the following components:

1. Access rights:
   a. Which employees should have access to what kinds and types of resources;
   b. When, where, and how often is access allowed through resources via the biometrics-based VPN;
2. Access control rights which include the following:
   a. IP address source;
   b. Data packet content and destination from within the biometrics-based VPN;
3. VPN management responsibilities, which include the following:
   a. Who will administer and oversee the biometrics-based VPN solution;
   b. Who will enforce the security of the biometrics-based VPN
   c. Who will authorize the issuance and the distribution of the digital certificates;
   d. Who will perform the registration activities.
4. The types and the degrees of the types and kinds as well as the degree of encryption which is required:
   a. Deciding upon the types of and kinds of IPSec protocol settings and options which are required;
   b. The management and distribution of the public keys and the private keys;
   c. The length of time for digital certificate activity and expiration;
5. VPN endpoints: This simply involves where the IP tunneling will be routed through:
   a. Gateway to gateway;
   b. Gateway to desktop;
   c. Desktop to desktop

In terms of a biometrics-based VPN System, all types of programs and applications which run in your place of business or organization need to be double checked in order to make sure that they will be compatible to the new VPN hardware and software. Such programs and applications need to closely examined, which include:

1. Database access and maintenance programs;
2. Mainframe access through terminal emulators;
3. Any type or kind of software development tools and their respective databases;
4. All sorts of dynamic Web content generators used for intranet development;
5. Any type or kind of document sharing program;
6. All remote server administration hardware/software;
7. All backup as well as remote backup tools utilized at your place of business or organization.

## End users and employees

With regards to the biometrics-based VPN System, the concern of users and employees should be of prime concern as well. As the CIO of your business or organization, the following end user requirements need to be taken into consideration before the biometrics-based VPN System is deployed:

1. The total number of remote office locations your place of business or organization is intending to fully support in a biometrics-based VPN system;
2. Where the existing corporate networks, such as the local area networks (LANs) and the wide area networks (WANs) cross and intersect with one another;
3. How many end users will need to be supported by the biometrics-based VPN System, and this includes regular employees, business partners, and contracted employees;
4. The type and kind of end users the biometrics-based VPN must support, such as the following:
   a. Software developers;
   b. Sales force employees;
   c. Facilities personnel;
   d. Office personnel;
   e. Customers;
   f. Contractors;
   g. Business partners;
   h. Suppliers/vendors

As the CIO, it is equally important to consider is the type and kind of access to the applications all of your end users will be needing, such as:

1. Access to particular network directories;
2. Access to all types and kinds of corporate and remote servers;
3. Access to all types and kinds of network resources;
4. Mainframe access;
5. Database access to the software;
6. Web based and HTTP support;
7. E-mail server support;
8. Client–server applications;
9. Mainframe terminal emulation.

## The network requirements

Above all, to the CIO, determining and ascertaining the biometrics-based VPN system network requirements is of prime importance above anything else and includes the following issues which need to be taken into consideration:

1. The design and topology of the network layout and structure;
2. The various access points into the overall corporate network and the Intranet;
3. Any type or kind of dynamic protocol support;
4. Internal protocol service requirements;
5. All and any existing routers, firewalls, and proxy servers which exist in both inside and outside of the place of business or organization;
6. All types and kinds of any existing authentication rules which exist in the place of business or organization;
7. Any and all business applications which will cross into the boundaries of biometrics-based VPNs;
8. All bandwidth requirements set forth by the place of business or organization;
9. Any cryptographic processing requirements;
10. Information technology staff and support requirements;
11. The full scalability of critical IT devices such as:
    a. Authentication servers;
    b. Database servers;
    c. Web servers;
    d. Any and all mapped driver and directories;
    e. Corporate e-mail servers and gateways;
    f. Corporate and end user FTP servers;
    g. Remote network /server administration.

Finally, as the CIO, evaluating the performance of the biometrics-based VPN is just of equally if not more of prime importance, and these items need to be reviewed such as:

1. The cryptographic hardware accelerator support;
2. The clustering of servers for scalability;
3. All types and kinds of quality of assurance service levels for all of the corporate servers, extranets, and intranets;
4. The importance of back up and redundant devices in order to ensure full and complete non-stop processing of the information and data for the place of business or organization;
5. Any expected growth in the network bandwidth requirements.

As the CIO of your place of business or organization, more than likely, you will be purchasing a biometrics-based VPN system solution which comes out of the box and is ready to install, perhaps even with the help of an outside consultant.

## Building your own virtual private network

But if you are one of those more daring types of CIOs who likes to perhaps implement and design your own biometrics-based VPN solution, there are a number of key things which you need to be aware of. Now of course, it is out of the realm of this chapter to go over completely what specifically needs to be taken into consideration when crafting your own biometrics-based VPN system, but a few critical points can be offered here at this juncture of the chapter.

First, as the CIO, it is very important to keep in mind that there are three levels of information technology systems that are greatly impacted by the addition of a biometrics-based VPN system. These three items are typically the web servers, the application servers, and the database servers.

But, this is also known as a three tier advantage, and possesses the following strengths:

1. Information technology is greatly enhanced;
2. The server scalability is much easier to implement;
3. The centralization concepts allows for the IT staff to control and to secure programs and servers;
4. Well defined software layers at the place of business or organization's software model allows for a quick responsiveness from the IT team from an ever changing business environment;
5. Any type or kind of existing mainframe services can be reused throughout over time;
6. Biometrics-based VPN systems based on open-source software model allows for the places of business and organizations to incorporate new technologies very quickly and readily.

## Impacts to the web server

If you are going to design your own biometrics-based VPN solution, then the first line of impact will be that of the web server, which can be seen in Figure 4.1. Although many CIO's have heard of the term "Web Server" before, there is special type of web server which merely responds to the Hyper Text Transport Protocol (HTTP) requests for the HTML pages it delivers to your customer's web browsers.

A key example of this would be your company catalog that your customers are trying to view online. To improve the response time of the web servers and to deliver high-end quality web pages to your customers, there are three types of processing which can be utilized, through a biometrics-based VPN system, and they are as follows:

1. Symmetric multiprocessing, or SMP: This is when a web server relies upon a single multiprocessor as well as a single memory capability in order to accomplish the workload which has been assigned to them, but the scalability of the SMP is a very serious limitation;
2. Massively parallel processing, or MPP: This is when the role of the logic chip has its very own processor as well as its own memory subsystem to process information and data. These are all threaded to the nodes, and collects the results on the other end. As one can tell, these types of processors are very expensive, but also offer limited scalability to the biometrics-based VPN system;
3. PC clustering: In a biometrics-based VPN system, this type of approach bridges the first two processing techniques as described up above. This type of system can contain unlimited subprocessors, all connected via a fiberlink, or what is known as an optical network. With this type of system, each SMP runs its own version of the operating system, and can provide strong backup features. But, the biggest drawback is that this type of system is that it is quite cost prohibitive.

## Impacts to the application server

The second level of Information Technology system which will be greatly impacted by the adaptation and the implementation of a biometrics-based VPN system is that of the Application Server. Simply put, these types and kinds of servers store, manage, operate, and control the specialized software components which are relevant and of course, very important to your place of business or organization.

When implementing a biometrics-based VPN system, as the CIO, you may also want to consider implementing a trusted host. This simply makes use of Access Control Lists (or ACL's) in an effort to thwart hacking attempts at running or installing various application and software programs, provided that they do not have the authority to be installed onto the Application Server.

It should be noted that access control information which are stored upon the Application Servers should be stored on as the most garbled form possible, and never stored as cleartext anywhere. As the CIO, you should remember not to place too much work on the type or kind of Application Server, as in turn, this could place a great strain on the biometrics-based VPN system, by adversely impacting the inflow of data packets.

## Impacts to the database server

Finally, the third level of Information Technology system which could be impacted by the implementation of the biometrics-based VPN system is that of the Database Server. For your place of business or organization, these store the product, the purchasing data, and the customer data as well as all of the other data warehousing techniques which are contained within it.

For the Database Servers, the Object Oriented Database Management System, or the traditional relational databases could be used. But regardless of this choice, these specific types of database softwares must be "prepared" for the implementation of a biometrics-based VPN system. When the biometrics-based VPN is finally installed, at your place of business or organization, it is absolutely imperative to make sure that all company information and data is encrypted across all levels, which includes the Field Level, the Row Level, the Table Level, as well as the entire database level.

## Impacts to the firewall

Finally, as the CIO of your place of business or organization, another key component which cannot be overlooked is that of the firewall-in fact, many security experts agree that putting the biometrics-based VPN in front of the firewall is one best choices that can be made. For example:

1. A single set of rules allows you, as the CIO, to control the flow of data packets in and out of the biometrics-based VPN;
2. A single point of administration helps you, the CIO, to manage changes quickly as your business or organization dictates;
3. Finally, by having a single point of access at the firewall level, this allows you the CIO, to secure and control your IT systems very responsibly as well as responsively.

Now, that we have looked upon some of the critical issues which are important to you, as the CIO in terms of some of the major design considerations if you are creating your own biometrics-based VPN solution, we now need to look at another key aspect of security-which is that of Testing of your biometrics-based VPN system.

## Virtual private network testing

It does not matter if you have bought it off the shelf, or had your IT team design it the whole thing from scratch. Once your biometrics-based VPN system is in place, as the CIO, you need to security test the whole system from a number of different angles, which include providing different security alternatives which include:

1. Attacks against biometrics-based VPN solutions include:
   a. Always destroy the plaintext;
   b. If you use what is known as virtual memory, but make sure that the plaintext which is stored in them always gets destroyed;
   c. From the biometrics-based VPN system, delete any end users passwords from memory after the dialog box closes;
   d. Make sure that the encrypted ciphertext are both protected with strong keys, not just a strong one and a weak one;
   e. Do not let the master keys be governed by the mercy of your session keys;
   f. As much as possible, make sure that every possible method in which a hacker can guess or learn both your public keys and private can be mitigated;
2. Attacks biometrics-based VPNs hardware can be attacked through the following methods:
   a. Measuring power consumption usage;
   b. The ability to measure radiation emissions from "other side channels" in the biometrics-based VPN;
   c. Fault Tolerance Analysis;
   d. Purposely and deliberately introducing faults in order to determine both the public keys and the private keys.
3. Attacks against trust modules-As much as you have to be aware against the technological threats to your biometrics-based VPN system, as the CIO, you have to be aware of the human threats which could take place, with the end resultant being that the biometrics-based VPN system is totally compromised. Such human threats include the following:
   a. Which personnel in the place of business or organization can be trusted?
   b. How does the biometrics-based VPN confer trust amongst trust of the IT staff?
   c. How far should this trust be extended?
   d. How do you, as the CIO, avoid the possibility of collusion between the end users of the biometrics-based VPN system?

e. As the CIO, what checking mechanisms can you put into place to make sure that human collusion safeguards are indeed being enforced upon?
f. How do you avoid the human temptation that you can fully 100% trust all of your IT staff at your place of business or organization?
g. Making sure that the version control to the software code changes made to the software applications of the biometrics-based VPN system is firmly locked down and secure;
h. Getting away of the notion of convenience versus security for the IT staff as well as the entire business or organization;
i. Making sure that your biometrics-based VPN system is as much 100% irreversible engineered as possible;
j. Having the false sense of security that a solid trust model is always going to be 100% fail-safe, and will never need to have any improvements made to it what so ever. Truth be told, improvements will always have to be made, and the security trust model implemented will always have to be updated to keep with the so-called cat and mouse game against the hackers (this simply means that being at least two to three steps ahead of the hackers as they devise new hacking techniques in just a matter of hours or even minutes);

4. Attacks against failure recovery: Failure recovery describes the system's capability to have the ability to maintain high levels of security in the face of any component breakdown or failure. In other words, to the CIO, this becomes the classical war of functionality versus security-in other words-that fine line has to be drawn and balanced of what is more important to your place of business or organization in the face catastrophic breakdown: Your customer's convenience or the businesses backup and recovery plans?
5. Attacks against cryptography: Yes, even despite having the best possible biometrics-based VPN system in place, and the best cryptographic tools to go with it, cryptographic flaws can and still do happen. For instance, to help protect against this, some of the following steps should be taken:
   a. Keep the mathematical values as random as possible, all the time;
   b. Make sure that the cryptographic digital signatures are strictly correct about the parameters they need and should be in;
   c. From the standpoint of the CIO, you should know and understand what the appropriate and the inappropriate uses of the biometrics-based VPN protocols;
   d. Establishing the biometrics-based VPN testbed: As a CIO, you must remember that your biometrics-based VPN must be able to communicate and process the data packets with any other

biometrics-based virtual private network systems not only at your place of business or organization, but at other business establishments who may also possess such a VPN. To allow for this free flow of communications between biometrics-based VPN systems, the communications become the corporate intranet and the Internet becomes of paramount importance. To accomplish this task:

- Make sure that you dedicate PC's for each individual client;
- Issue test versions of the X.509 digital certificate for each end user and each network resource requiring access to the biometrics-based VPN system;
- Use trusted Internet Service Providers to allow for access form the remote locations, especially where remote connectivity is required;
- If possible, exchange a few of the digital certificates to the test population group at your place of business or organization;
- Never, ever use your own corporate intranet for the testing of the biometrics-based VPN system between the client PCs;
- Seriously and strongly consider putting test machines to test your biometrics-based VPN system at different, offsite locations;
- If possible, as the CIO, try to use some test cases for both the network gateways and the clients. This will enable you and your IT staff to compare the test results equally between the various biometrics-based VPN systems;
- Most importantly, as the CIO, document your test results from the initial testing of the biometrics-based VPN system, and report any types and kinds of problems, malfunctions, and abnormalities to the vendors who have manufactured the requisite hardware and software, so that the problem can be rectified and mitigated as quickly as possible.

## Implementing a virtual private network

Now that we have looked at some of the major design implementations of a biometrics-based VPN system, the next step is to is its proper implementation at your place of business or organization.

Implementing your biometrics-based VPN system, you as the CIO, can craft an easy four step plan implementation plan:

1. Development of the VPN security plan;
2. Preparation for the actual installation;
3. Performance of the installation activities;
4. The biometrics-based VPN system verification.

In terms of the development of the security plan, you need to, as the CIO, carefully craft and design the policies, the procedures, and the rules which govern access to all of the network resources which include:

1. The host computers and the servers;
2. The computer workstations;
3. The various network gateways, routers, firewalls, and network bridges;
4. The terminal servers and the remote access servers;
5. The network operating software;
6. The operating system software of all of the servers;
7. Any type and kind of server application software which may be in use;
8. All and any types of mission critical data found at your place of business or organization.

But keep in mind, as the CIO, building upon the security plan must include those end users who will be the most affected by the implementation of the biometrics-based VPN system. As the CIO, you need to carefully formulate the types and kinds of services that your end users will need from the system, especially getting permission rules for the user groups which are to be established.

Also, with regards to the VPN system rules for your end users, there must be extremely tight security, both physically (in terms of access control) and logically (granting who has network access to the biometrics-based VPN system).

Equally important to the protection of the biometrics-based VPN system is that of the physical protection of both the public keys and the private keys. Also, as part of developing the security plan for the biometrics-based VPN system, you need to collect information and data on the following types of hardware which exist in your Information Technology infrastructure:

1. The external proxy servers;
2. All of the Internet Routers;
3. All types and kinds of firewalls as your place of business or organization;
4. All of the virtual servers.

In addition to the above, as the CIO, you will need to set up access control policies for the following items:

1. All of the VPN-based domain name service host names and the IP resolution based tables;
2. All of the e-mail network protocols used at your place of business or organization;

3. The web servers in use (such as internet information services for Microsoft environments and Apache for Linux environments);
4. Any Telnet and File Transfer Protocol networks and connections;
5. Any other kinds and types of socket protocols and customized network protocols.

Once the above is done and accomplished, as the CIO, you then need to define the user groups which will exist at your place of business or organization. This step simply involves the building and establishing end user tables, as well as defining the access roles and rights of each group.

One key benefit of establishing these groups is that you can create a single entity which will fit all of the needs of each end user, rather than having to create rights and permissions for each individual employee at your place of business or organization.

The next phase is to properly define your biometrics-based VPN System Tunnels, which include the following:

1. Remote user access to the biometrics-based VPN Gateway;
2. Gateway to gateway exchanges and connections;
3. Gateway to remote partner exchanges and connections.

Now finally, you can get your biometrics-based VPN system set up for the final installation. This involves a full assessment of all of the following critical resources at your Information Technology Infrastructure:

1. The operating system readiness;
2. All dedicated hardware properly configured;
3. Physical access control checks to the VPN system are set in place and motion;
4. All of the network hardware and software is fully set to go;
5. All of the network routing tables to the biometrics-based VPN system are established and fully verified;
6. All of the host names at your place of business or organization match all of the domain name server entry names;
7. All of the TCP/IP addresses which relate to the biometrics-based VPN system are known, established, and operating properly and correctly.

## Managing public and private key exchanges

One very important aspect to remember about installing a biometrics-based VPN system is that of key exchanges. Managing key exchanges can come in three different methodologies:

1. Manual key exchanges: With this system, a VPN administrator must configure each biometrics-based VPN system with its own set of public keys and private keys, and also with the public keys and the private keys of the other biometrics-based VPN systems so that they can all communicate effectively and efficiently with each other. The prime disadvantage with this methodology is the lack of scalability;
2. The simple key interchange protocol: This is also known as "SKIP". This protocol is only available for highly commercialized applications and implementations of biometrics-based VPN systems;
3. The ISAKAMP/Oakley protocol: ISAKAMP is the de facto standard for the IPv6 network VPN protocol, and creates the framework for both the public key and the private key management. This type of VPN protocol does not actually the public keys and the private keys, rather it is used with the Oakley Network Protocol to produce and establish the public keys and the private keys for the biometrics-based VPN system.

As the CIO, a key principle to help fortify the security for the public key and private key management of your biometrics-based VPN system is to centrally administer the remote VPN Gateways, and ensure that the interface to the biometrics-based VPN software is not available on any other network interface or device.

As it was eluded to previously in this chapter, digital certificates are a very important component of Cryptography. But not only it also plays a very crucial role with the biometrics-based VPN system.

Since the biometrics-based VPN's primary function is to filter in and out the good and known data packets, these same types and kinds of digital certificates can be used to control access to all of the primary servers at your place of business or organization.

## The access control list

This is done primarily via the ACLs. Essentially, this correctly identifies the list of approved end users along with their passwords at your place of business or organization.

Digital certificates can be used as a very strong proxy for each of the employee's username and password pairs. To make this happen, the digital certificates of each end user/employee associated to the ACL is mapped, and can even determine and ascertain the access rights by reading each of the digital certificate's specific contents.

To make the above process occur for your biometrics-based VPN system:

1. Have the ACL direct each new employee and end user to a brand new digital certificate for each separate site access;

2. The new employee/end user may have to select a new User ID / Password combination;
3. The new employee/end user can then log back into the biometrics-based VPN system, and the digital certificates is mapped to the new employee's/end user's User ID and Password combination.

But as the CIO, as the access rights of your new employees change, you for sure will want to have to avoid the situation where new digital certificates have re-issued each time that the rights and privileges of your employees and end users change each time, per their job requirements and duties.

To do this, you, as the CIO, need to use a X.509 secret key to help secure your biometrics-based VPN system. To accomplish this task, follow this specific methodology:

1. Generate a random X.509 private key;
2. Because the digital certificate is already embedded into the biometrics-based VPN system, a challenge/response system is created;
3. Once the challenge/response has been successfully verified through the software at your Information Technology infrastructure, it can then create its own challenge, and automatically sign the end result, which will then use the private key associated with the X.509 digital certificate;
4. The signed result end user certificate is then sent back to the biometrics-based VPN server;
5. The verification of the end user's certificate is then performed;
6. If the digital certificate is proven to be valid, the end user's challenge is then checked, utilizing the public key in employee's or end user's own digital certificate;
7. If the digital certificate is proven once again to be valid, the resultant challenge is checked against the original challenge in order to ensure that they match together;
8. If this particular challenge matches all of the previous steps just outlined above, the digital certificate number the information and data can then be extracted from the digital certificate, and utilized as a lookup into the ACLs which contain those particular employee/end user access rights and privileges.

As one can see, as a result of the recent advent of the biometrics-based VPN systems, a number of key security drafts, over time, have created around the security issues of VPN's, and these drafts are also as the "Internet Drafts".

## Internet drafts

One such draft which relates to this chapter is known as the "Internet X.509 Public Key Infrastructure". Specifically, this relates between the Elliptical

Digital Certificate Algorithm (this is also known as Elliptical Wave Theory, and was also detailed earlier in this chapter) in the PKI X.509 Certificate schemes.

Here are the specific subjects, broadly speaking, this Draft covers:

1. Certificate management messages over CMS:
   This generally specifies the need for a Certificate Management Protocol utilizing CMS in two different ways:
   a. The need for an interface which connects to the PKI products and services;
   b. This outlines the needs for a certificate enrollment protocol using the Diffie–Hellman public keys (which was also reviewed in detail earlier in this chapter);
2. The Internet X.509 PKI time stamp protocol:
   Time stamps prove beyond a doubt that a transaction took place at a certain point in time, and thus it serve as a "Trusted Third Party", or TTP for short. This piece of the Draft regulates how the format of the timestamp request should be sent to a Time Stamping Authority, and what kind and type of particular response is valid;
3. The X.509 PKI data certification server protocols:
   This document describes very specifically the data certification services, and the network protocols used when communicating with it:
   a. The validation of the digital signature;
   b. Providing updated information with regards to the status of the public key certificates;
4. The Internet X.509 PKI PKIX roadmap:
   This document goes into the actual detail of the theory and the implementation of the X.509 digital certificate based PKI;
5. The Internet X.509 PKI qualified certificates:
   This document specifies the digital certificate profile for what is known as Qualified Digital Certificates. The primary purpose of this document is to define the syntax of the language which shies away from all types and kinds of legal jargon.
6. The Diffie–Hellman proof of possession mathematical algorithms:
   This specialized document details the particular of two separate methodologies to produce a digital signature from a Diffie–Hellman key pair. The mathematical algorithms are designed to create an audit trail rather than just for digital signing;
7. The X.509 attribute certificates:
   This document specifies two very mathematical base profiles, which are used to provide digital certificate authentication services;
8. The basic representation token V1:
   This technical document helps to define data structures, which have been established and distributed by a trusted provider;

9. The Internet X.509 PKI extending trust:
   This technical document establishes the technical details that are needed to maintain a certain level of confidence by a trusted provider;
10. The Internet X.509 PKI operational protocols, LDAP V.3:
    This document highlights in details the LDAP V.3 features which are required to maintain a PKI based upon the X.509 digital certificates;
11. The simple certificate validation protocols, also known as SCVP:
    This specialized network protocol permits the client computer to allow the primary server to handle the digital certificate processing. As a result, this type of server can then provide certain types and kinds of information and data about the details of the digital certificates, especially its validity or invalidity.

## Four vulnerabilities of a biometrics virtual private network

With regards to the biometrics-based VPN, there are four critical areas where the data packets, which house the biometric templates are at most risk to hacking and theft, and they are as follows:

1. Just after template creation (this includes both the verification and the enrollment templates);
2. The biometric templates which are housed in the database (the actual database depends upon the specific biometric technology being used);
3. In client–server network topology, the transmission of biometric templates from the biometric system to the central server (this is where the biometric database resides);
4. In a hosted environment, there the biometric template database resides with a third party.

## Biocryptography

Biocryptography provides the means to further biometric templates at these critical junctures. As it was reviewed at the beginning of this chapter, Cryptography is the science of scrambling information and data which is transit across a network medium, and then descrambling it at the receiving end into a decipherable format.

That way, if the scrambled information and data were to be intercepted by a third party, there is not much which can be done unless they possess the keys for descrambling the information. These concepts of scrambling and descrambling can be very easily applied to biometrics. This is formally known as "Bio-Cryptography".

In other words, the Biometric Templates are protected by scrambling and descrambling keys while they are stored in the database, or in movement across a network. To further illustrate the exact nature of Biometric Templates, two types will be examined: Fingerprint Recognition and Iris Recognition templates.

To review, whenever we send a message to our intended recipient-whether it is by e-mail, instant message, or even just a text message on our Smartphone, this message is often sent as a "plaintext", or "cleartext".

This means that the actual message is being transmitted to the intended recipient in the way it was originally constructed by the originator of the message. Thus, the only true way to protect the information being sent is to scramble it, in other words, "encrypt the message". This encrypted (or now scrambled message) is now known as the "ciphertext".

The reverse of this process is known as "decryption", with the end result being a readable message to the intended recipient. As all of this relates to biometrics, the data packet which houses the biometric template (such as the fingerprint or iris recognition template) can be viewed as the plaint text, or as the "Plaintext Biometric Template".

## The cipher biometric template

When the fingerprint or iris template is encrypted, it can be viewed as the "Cipher Biometric Template", and when it is decrypted, it can be viewed once again as the decrypted "Plaintext Biometric Template". But other than just doing the above, Biocryptography also has to provide the following three functions in order for it to be truly effective:

1. Authentication: The receiver of the message (or the Plaintext Biometric Template) should be able to, 100%, verify the origin of it;
2. Integrity: The message in transit (or the Plaintext Biometric Template) should not be modified in any way or format while it is in transit (or in other words, replacing a fingerprint biometric template with an iris biometric template in order to spoof the biometric system);
3. Non-repudiation: The sender of the Plaintext Biometric Template should not falsely deny that they have not sent that particular template originally.

## Biocryptography keys

A component which is central to biocryptography are what is known as "keys". It is the key itself which is used to lock up the Plaintext Biometric Template (or encrypt it) at the point of origination, and it is also used to unlock that same Template at the receiving end.

The key itself is a series of mathematical values-obviously, the larger the value, it is equally (if not more) that much harder to break it while it is in transit. The range of possible mathematical values is referred to as the "keyspace".

There are many types of such keys used in biocryptography, such as signing keys, authentication keys, data encryption keys, session keys, etc. The number of keys which are generated depends primarily upon the mathematical algorithms which are used, which are primarily symmetric and asymmetric mathematical algorithms.

To further fortify the strengths of a biocryptography based PKI, mathematical hashing functions are also used to protect the integrity of the Plaintext Biometric Template. For example, when the destination party receives the Plaintext Biometric Template, the hashing function is included with it.

If the values within the hashing function have not changed after it is has been computed by the receiving end, then one can be assured that the Plaintext Biometric Template has not been changed or altered in any way.

To prove the validity of the hashing functions, it should be noted that they can be calculated only in one direction (e.g. going from the sending point to the receiving point, where it is computed), but not in the other direction (e.g. going from the destination point to the origination point).

## A review of how biocryptography can be used to further protect fingerprint and iris templates

As it was discussed in Chapter 3 of this book, biometric systems can take quite an array of system designs, deployments, and network architectures. To illustrate the concepts of Cryptography can be used with biometrics, three primary examples will be reviewed:

1. From the standpoint of a single Biometric System;
2. A client–server setting, where the biometric devices are connected to a central sever;
3. A hosted environment, where the biometric templates and the processing functions are placed in the hands of a 3rd party.

## Biocryptography in a single biometric system

With the biometric technology which is available today, most fingerprint recognition and iris recognition scanners consist of the database and the processing functions in just one unit. Meaning, enrollment and verification occur at a single point.

Obviously, there are many advantages to having this type of "stand alone" system, with the two biggest ones being low overhead in terms of costs; and very quick times in terms of processing the enrollment and verification templates.

This is how biocryptography would be used to protect the iris and fingerprint recognition templates in this type of environment:

1. Assuming that the end user wishes to gain physical access or logical access with either their fingerprint or their iris scan, a verification template must be first created. This template would either be a binary mathematical file (for a fingerprint template) or an IrisCode (for an iris template);
2. Using the principles of symmetric cryptography, the verification templates would then be encrypted with the key that is generated by the system. This would occur right after the unique features from the fingerprint or the iris are extracted, and the template is created.
3. Once the verification template reaches the level of the database, it would then be decrypted by the same key, and the statistical correlations would then be computed between the verification and the enrollment templates. If this correlation is within the bounds of the security threshold established, then either physical access or logical access would be granted by the biometric system.

In this illustration of the stand alone iris or fingerprint recognition system, a number of key assumptions are made:

1. Only verification, or a 1:1 match is being used. And since it is being done at the local level, the configuration needs are low, thus symmetric algorithms is the key choice of Cryptography to be used. As a result, only one key is generated.
2. Only the verification templates receive the added protection from encryption. This is because the fingerprint and iris recognition templates are created only once, and later discarded from the respective biometric system. In fact, in any biometric system, no matter what the magnitude of the application is, verification templates are used only once, and no more.
3. The enrollment templates in the finger and iris biometric system receive no added protection from encryption. There is no doubt that this is an inherent security risk, but one has to keep in mind also that the database is being stored in the biometric device, and not located in multiple places, where the need for protection is much greater.
4. In a biometric verification application at the local level, the processing power required is much less than when compared to the client–server, or even the hosted approach. Thus, this further supports the need for only the use symmetric algorithms.

## Biocryptography in a client–server biometric system

In a client–server network topology in the traditional sense, there are a series of computers connected to a central server, via a network medium (for example, it could be a hard wired network, or even wireless).

From within this central sever, resides all of the resources and applications the end user needs to access to. Not only this, but databases are very often stored here, which contain all types and kinds of data, but this is also the point where database querying and processing takes place.

This type of infrastructure can be very small (also known as a LAN), or it can be very large covering great distances and international boundaries (this is known as a WAN).

This type of setup can also be extrapolated to biometrics. Multiple biometric devices can be networked to a central server, in a very similar fashion as described above. But the key difference in a biometrics client–server system is that the primary application is for verification and identification, template processing/querying, and nothing more.

Today, biometric client–server applications do exist in business today. Typically, it is the medium to larger sized businesses which have this type of configuration, because many more resources are required, which can be a much greater expense than versus the single biometric system, discussed previously.

In this type of configuration, the most commonly used biometric devices are that of hand geometry scanners, fingerprint scanners, iris scanners, and even facial recognition scanners. Different biometric devices can be used (such as fingerprint scanners being used in conjunction with iris scanners), or the same biometric devices (for example, all fingerprint scanners).

In the end, it really does not matter what hardware is used, because they are all accessing the same resource-namely, the central server. But with all of this, comes an even greater risk-specifically the security threats which are posed to the system.

Again, biocryptography can play a huge part to insure the protection of the biometric templates. For this configuration, asymmetric cryptography will be used. Meaning, a public key will be generated, as well as a private key.

This is how biocryptography will be used in this type of environment:

1. The end user either has their finger or their iris scanned, and the usual verification templates are created by each biometric system (assuming that both iris and fingerprint scanners are being used simultaneously);
2. These iris and fingerprint verification templates will then be protected by a key, specifically, the public key;
3. After the above two processes have been completed, the newly encrypted verification templates will then make its way across the

appropriate network media, and finally make their way to the central server;
4. It is here where the biometrics database is stored, which contains the relevant iris and fingerprint enrollment templates, and it is here where the private keys are stored and also encrypt these templates;
5. Once the public and private keys have been decrypted, the appropriate statistical measures will be applied to determine the closeness or the dissimilarity between the verification and the enrollment templates;
6. Based upon the results, the end user will be granted or denied access to whatever application they are trying to gain access to (either physical access entry or logical access).

An important point need to be made about this biometrics client–server system:

1. As is the case in the single biometric system, the verification templates, after they have been compared and evaluated, will be discarded along with their public keys. The enrollment templates will have to be decrypted as well, so that the comparison between the verification and the enrollment templates can be made. But, because the enrollment templates are stored from outside the actual biometric device, they will have to be encrypted again to insure maximum security while they are stored in the database at the central server. Thus, this type of configuration will need far more network resources and much more processing power. Also, there is no doubt that extra overhead and quite possibly further verification times will increase by a few more seconds (under normal conditions, this happens in less than one second).

## Biocryptography in a hosted biometrics environment

The world of information technology is now moving very quickly towards a new type of application-which is known as "Cloud Computing", or "Software as a Service (SaaS)". This means that a businesses' entire IT infrastructure can be managed and outsourced to an independent, 3rd party, also known as a "Hosting Provider".

At this venue, all of the IT infrastructure hardware and software is set up and managed, maintained, and upgraded by them. All a business owner has to do is open an account with this hosting provider, and with a few clicks of the mouse, set up the IT services they need or desire to have.

The primary advantages of this are: (1) Total elimination of IT administrative headaches and hassles, because the hosting party is held entirely responsible for all of this; and (2) A business only pays for the software/hardware services they have subscribed, at a fixed monthly cost.

This model of Cloud Computing can now even be expanded to the world of biometrics, which can be termed as "Biometrics as a Service (BaaS)". This is also known as a "Hosted Biometrics Environment". It would be established as follows:

1. All that a business would have to do is purchase and acquire the requisite biometrics hardware, in this scenario, it would be the fingerprint scanners and iris scanners;
2. All of the servers, the databases which house the iris and fingerprint enrollment templates, the processing of the verification templates between the enrollment templates, and formulating the match/non-match result rest entirely with the hosting provider. Meaning, after the biometrics hardware is installed, all the business has to do is simply set up the services needed with their account, and all is set to go.

But, because the iris and fingerprint templates are placed at the hands of the hosting provider, security of these templates obviously becomes the prime concern.

## Biocryptography and virtual private networks

There is yet another tool of Cryptography which would perfectly here – this is known as a "Virtual Private Network", or "VPN" for short, as just previously reviewed in extensive detail. This is how it work with a BaaS type of application:

1. The end user in the previous examples of this article has their fingerprint or iris scanned to create the respective verification template. This template then gets broken down into a separate data packet. This data packet (which contains the iris or fingerprint verification template) is then further encapsulated (or encrypted) into another data packet, so it eventually becomes invisible as it traverses across the various network media, as it makes its way to the servers of the hosting provider;
2. To ensure the integrity of the this double layered data packet, it would also consist of headers which contain information about the size and type of the verification template. This would be a confirmation to the hosting provider, as they receive this data packet, that none of the iris or fingerprint template data has been altered or changed en route;
3. To create another layer of protection, a dedicated VPN channel can be created. This would be a direct line from the point of origin

of the fingerprint or iris scanner all the way to the servers of the hosting party. This is known as specifically as "IP Tunneling", and as a result, this channel cannot be seen by other people accessing the Internet across the same network media the data packets (which contains the verification templates) are also traveling across;

4. Once this data packet arrives at the servers of the hosting provider, it is then decrypted, and the same processes of verification/enrollment template comparing, as detailed in the biometric client–server network system, will be followed;
5. After a match or non-match has been determined, the result is then sent from the hosting provider back to the place of origin of the iris or fingerprint scanner, and the end user is then allowed or not allowed access to the resources or applications they have requested.

## IPSec

To further fortify the VPN between the place of business and the hosting provider, a protocol known as "IPSec" can be used. The IPSec protocol is a major security enhancement to the TCP/IP protocols which used by everybody today to gain access to the Internet.

IPSec also uses digital certificates (which are also the public and the private keys). It should be noted that there is one specific type of IPSec mode which help to provide the maximum security possible to the data packet housing the verification template.

It is known as "IPSec Tunneling", in which the header and the payload it is carrying (which is again, the iris and the fingerprint verification template) are further encrypted at a much deeper level.

Some special points need to be made about BaaS:

1. Although this type of application is still very new (only voice biometrics thus far has been used as a BaaS), it could very well be the wave of the future. This is so because, biometrics as a security technology, is still perceived to be very expensive. But with a hosted approach, these high costs will be greatly eliminated, because all that is required of the business owner to purchase is the required biometric hardware, and also, like SaaS, BaaS would a fixed monthly cost.
2. With BaaS, the verification times and presenting the match/non-match result could take time, because all of the template processing and matching will take place at the hosting provider. The speed will be a direct function of the hardware and the software that is used, as well as the amount of network bandwidth that is consumed.
3. Although BaaS holds great promise, one of the biggest obstacles which it will face is one which has constantly plagued the entire

biometrics industry – which is the issue of Privacy Rights. This issue will only be proliferated more as the biometric templates are in the hands of an outside 3rd party, which is the hosting provider.

In summary, this chapter has examined how the principles of cryptography and biometrics can be used together to provide maximum security for the most important component – namely the biometric templates which are created and stored. As of this writing, biocryptography is still very much an emerging field, and some observations have to be noted:

1. The three examples described and discussed in this article are theoretical in nature-they have not yet been proven in the real world, and it is assumed that much research will have to be conducted before these applications are put to the test on a commercial basis.
2. These examples, if proven to be viable in the real world, could become very complex, very quickly. This is so because there are two types of security technology becoming fused as one, and from that, many variations of all kinds and types could be created. Thus, great importance has to be paid to the actual biometric systems analysis and design right from the outset, in order to help ensure a smooth streamlined process, from the standpoint of troubleshooting and support.
3. Another area which as haunted and plagued the biometrics industry is that of a lack or sheer absence of standards and best practices for the technology both in its current state and as it is being developed. As Bio-Cryptography emerges into the forefront of security, this is an absolute must, in order to avoid duplication of efforts, resulting in unneeded and bloated overhead. Also, as businesses and entities start to adopt Bio-Cryptography, a standards and best practices list will help to provide the groundwork which is needed to create security applications, rather than having to reinvent the wheel every time.
4. Finally, it is the author's point of view that Bio-Cryptography be developed in an open model type of forum, where all parties involved, ranging from the private sector to academia to the level of the government can collaborate and discuss new ideas with the goal of open communications and open dialogue. An open model would help to minimize the any potential security threats and risks to Bio-Cryptography, because answers and solutions can be very quickly created. This is best exemplified by the use of the open-source model for software development versus the closed source model.

## REVIEW

Our last chapter of this book looked at and reviewed the emerging field of biocryptography in extensive detail. An extensive overview of Cryptography was provided, as well as how biometrics could very easily fit into this scheme to help further fortify and strengthen the biometric templates which are created. In summary, some of the major components reviewed were symmetric and asymmetric cryptography, the components of PKI, as well as VPNs.

As was demonstrated in the last chapter, it appears that VPNs will be the way to go, if you will, for the world of biometric technology. Essentially this, simply put, involves encapsulating the data packet in which the biometric template resides in with another data packet. This particular data packet then traverses across a network connection which is virtually invisible to the outside world.

As it was also reviewed, all of the mathematical algorithms involved in biocryptography were examined, and it appears that the RSA algorithm will provide the most robustness for biocryptography, in a virtual private setting. This is so because the RSA algorithm relies upon the power of very large prime numbers in order "crack its code".

Chapter 5

# Biometrics in the cloud

## INTRODUCTION TO CLOUD COMPUTING

As we now move to the next chapter of this book, as well as its last, we now cover a very important area biometrics will play a role again is in that of cloud computing. Cloud computing has become a very popular phrase in the 21st century, especially to that of Corporate America. To many people, the only thing that is truly understood about it is that it brings much lower, fixed costs to a business owner at a predictable price. And also, it is popularly understood that an entire IT infrastructure can be placed at the hands of a third party, known as the internet service provider (ISP).

But truth be told, there is far more to cloud computing than just the above. And in fact, cloud computing is still a very nebulous concept. Therefore, the goal of this final chapter of the book is to dispel the myths surrounding cloud computing, and to provide the CIO with a strong background into it, and above all, how biometrics can fit into and play a role within the cloud computing infrastructure. Therefore, this last chapter will cover the following topics:

1. The basic concepts and terminology surrounding cloud computing;
2. The benefits of cloud computing;
3. The challenges and risks of cloud computing;
4. The functions and characteristics of cloud computing;
5. Cloud computing delivery models;
6. Cloud computing deployment models;
7. The security threats posed to cloud computing;
8. Important mechanisms of cloud computing;
9. Cloud computing cost metrics and service quality mechanisms;
10. An introduction to biometrics in the cloud;
11. A detailed example of how biometrics in the cloud will look like;
12. The advantages and the disadvantages of biometrics in the cloud.

 DOI: 10.4324/9781003562146-5

## THE BASIC CONCEPTS AND TERMINOLOGY SURROUNDING CLOUD COMPUTING

When one thinks of cloud computing, especially to the business owner, people get very excited. Why is this so? As eluded to in the introduction of this chapter, it is the buzzword which is floating around today, and will be for quite some time to come. Many business owners feel that if they adopt a cloud computing infrastructure, that they will be ahead of their competitors. While this may be true to a certain extent, the chances are that the competition has already beaten them to the punch.

This is so because in theory at least, cloud computing can be configured and started with just a few clicks of the mouse, and at the fraction of the cost outright owning an entire IT infrastructure. So, before we go any further into the components of cloud computing, just exactly how is cloud computing defined?

According to the National Institutes of Standards and Technology (NIS), cloud computing can be defined as

> a model for enabling ubiquitous, convenient, hands on, on demand network access to a shared pool of configurable computing resources (e.g. networks, servers, storage, applications, and services) that can be rapidly provisioned and released with minimal management effort or service provider interaction. This cloud model is composed of five essential characteristics, three service models, and four deployment models.
>
> *Source: Cloud Computing: Concepts, Technology & Architecture, Erl, T., Arcitura Education (2013, p. 28)*

This no doubt seems like a long and complicated definition, we now provide a much more distilled definition of cloud computing: "Cloud Computing is a specialized form of distributed computing that introduces utilization models for remotely provisioning scalable and measured resources" (Source: *Cloud Computing: Concepts, Technology & Architecture*, Erl, T., Arcitura Education, 2013, p. 28). An example of cloud computing can be seen in Figure 5.1.

Now, as we take both of these definitions, we find that cloud computing consists of a number of key components, which include the following:

1. The cloud;
2. The IT resource;
3. On premise;
4. Scalability.

*Figure 5.1* This is an image of a cloud computing environment. (Data center background cloud computing icon AI-generated image 2513262523 | Shutterstock.)

## The cloud

As we progress through these major components, the first one we look at is the cloud itself. This component can be specifically referred to as a "distinct IT environment that is designed for the purpose of remotely provisioning scalable and measured IT resources" (Source: *Cloud Computing: Concepts, Technology & Architecture*, Erl, T., Arcitura Education, 2013, p. 28). As we explore this component, the cloud is like a nebula in which various types and kinds of computing resources can be accessed.

These resources include the entire gamut of computing applications, which includes everything from the database to the individual software packages which service an entire small business, to accessing the File transfer protocol server, to accessing your e-mail server, and to even having the capability to develop and create complex software applications, to accessing content management systems for the corporate intranet, to even creating and launching an entire e-commerce platform.

## Two distinctions

But at this point, it is very important that two key distinctions be made. The first very important is that these computing resources just described are not accessed from a local hard drive on an employee's workstation or laptop. Or for that matter, these resources are not even accessed from a Local Area Network which can reside from inside a corporation. Rather, it is important to keep in mind that that these computing resources can be accessed from literally a thousand miles away, or even across the globe. The remote access is all made possible because of the Internet. With any standard web browser which is available today, any type or kind of cloud computing resource can be accessed.

The second very important distinction to be made is that all of these computing resources reside within an independent third party, which is privately owned, such as that of ISP.

## The IT resource

The second major component we now examine is that of the of the typical IT resources which can be found from within a cloud computing infrastructure. An IT resource, or even an IT asset, can be defined "as a physical or virtual IT-related artifact that can be either software-based, such as a virtual server or a custom software program, or hardware-based such as a physical server or a network device" (Source: *Cloud Computing: Concepts, Technology & Architecture*, Erl, T., Arcitura Education, 2013, p. 34).

Based upon this definition, and IT resource/IT asset is a tangible item which can be placed into the cloud, even some of the software applications which were just previously described. But, in the world of cloud computing, the IT resource/IT asset is typically a server, whether it is a stand alone physical server or just a virtual server. As it will be explored later in this chapter, a virtual server is also a physical server, but is partitioned so that each of the resources it hosts looks like it resides in a separate server, to the end user or small business owner.

So, as you can see, the most important IT resource, based upon the definition, is that of the server. After all, the software applications just previously illustrated need a place to reside in, or in cloud computing terminology, need a place to be "hosted" at. But, a key distinction must be made here. Although the virtual server can be placed into the cloud-it also represents all of the resources which are available to the end user.

A key feature, as it will be explained later in this chapter, is that cloud computing relies heavily upon shared resources in order to bring the low and predictable monthly costs to the end user and the small business owner.

## On premise

The third major aspect to be examined in the cloud computing infrastructure is that of the "on-premise" component. Specifically, on-premise can be defined as "being on the premises of a controlled IT environment which is not cloud based" (Source: *Cloud Computing: Concepts, Technology & Architecture*, Erl, T., Arcitura Education, 2013, p. 36). A key part of this definition is that of the "controlled IT environment". This environment is actually that of the ISP, or for that matter, any other private entity which has the flexibility to provide at least one or more IT assets and IT resources.

## Scalability

An important point needs to be made at this point is that the cloud-based resources are not just stored onto one specific computer, or even just one particular server. Rather, all of those IT assets and IT resources it consists of are shared across hundreds, or even thousands of servers at ISPs all over the world.

As we are now starting to see emerge, the cloud consist of a controlled environment, which is that of the ISP. The ISP consists of the IT resources and the IT assets which reside on a virtual server, from which these particular resources and assets can be accessed by the end user or the small business owner.

This is just one example of a general cloud computing infrastructure, and there is not just one of them. There are hundreds and thousands of them worldwide, to bring to the end user and the small business owner the shared computing resources they need and require. As it will emerge throughout the rest of this chapter, the cloud possesses many strong benefits, which are available to the end user and small business owner at an extremely affordable price, when compared to the other IT assets and IT resources which are available in the traditional models.

And, this will be especially true when biometrics becomes more available via the cloud infrastructure. On the spectrum of security technologies which are available today, biometrics technology tends to be more on the expensive side in terms for the customer with regards to procurement and deployment. But, with the cloud resources which are available today, and well into the future, an enterprise grade level biometrics technology infrastructure will soon be able to be procured and deployed by even the smallest of businesses and organizations, which at the current time, can only be afforded by the large corporations.

But apart from costs, the cloud infrastructure possesses other key benefits such as:

1. IT asset scaling;
2. Greatly reduced investments and proportional costs;
3. Scalability;
4. Increased IT availability.

## Asset scaling

In terms of the first major benefit, IT asset scaling merely refers to the fact that an IT asset or even IT resource can handle increased or decreased end user demand usage. There are two types of scaling which are available:

1. Horizontal scaling;
2. Vertical scaling.

With horizontal scaling, this is the allocation or the provisioning of valuable IT resources and IT assets which are of the exact same nature and type. The allocation of these particular IT assets and IT resources is known as "scaling out", or the releasing, or the giving away of IT resources to newer end users and newer small business owner is known as "scaling in".

For example, let us illustrate this with a well known application from Microsoft, known as SQL Server. This database software is widely available through most ISPs (also known as "Cloud Providers" in some circles). The allocation of SQL Server to the existing end users of this particular application is known as scaling out. But, the releasing of the SQL Server application to the new end users or the new small business owners when they first sign up for cloud-based services is a perfect example of scaling in. This type of scenario is very typical amongst the ISPs that provide the cloud-based IT services.

With regards to vertical scaling, the valuable IT assets and IT resources are merely replaced with a lower or higher version of the very exact same type or kind of IT asset or IT resource. For instance, replacing an IT resource or an IT asset with a higher version is known as "scaling up", and the opposite, which is the replacement of an IT asset or IT resource of a lower version is known as "scaling down".

Let us illustrate again with the same SQL server example, but this time, with a known version of SQL Server 2012. The offering to the end user a lower version of SQL Server (such as SQL Server 2000) is known as scaling down, and the offering to an end user a later version of SQL Server 2012 is known as scaling up. This type of scaling is less commonly offered by the ISPs as opposed to the horizontal scaling.

This is so because vertical scaling requires more down time. Vertical scaling typically only happens in the cloud environment when a later version of an IT asset or IT resource becomes available and makes the present IT asset or IT asset totally obsolete and outdated. With respect to the second major benefit of cloud computing, which is that of Reduced Investments and Proportional Costs, it is quite true that many of the ISPs that offer cloud computing services to their end users can have a revenue model based upon the bulk acquirement and acquisition of the IT resources and IT assets, and in turn, offer these services to their end user customer base at very cheap price point packages.

This is what directly allows for the owners of small- to medium-sized businesses to purchase a Fortune 100 enterprise grade level IT infrastructure for virtually pennies on the dollar. This is obviously of great appeal to the small business owner, as they can now replace capital operational costs and expenditures (which also means the total investment and ownership in an outright, entire IT infrastructure) with a proportional cost expenditure (which means the rental in a cloud-based IT infrastructure).

Because of the benefits provided by the proportional cost expenditures, this permits the small- to medium-sized enterprises to reinvest the saved money into other much needed IT assets and IT resources in order to support mission critical business functions.

But, keep in mind that ISPs are also businesses who have revenue generating models, and in order to afford such mass acquisition of IT assets and IT resources, they need to find inexpensive alternatives also. So thus, the ISP locates both types in data centers where the two most important cost variables of network bandwidth and real estate are the cheapest possible.

## Proportional costs

In this second category of a major benefit, there are a number of sub-benefits of cloud computing which are also available to end users and small business owners:

1. The access to on demand IT resources and IT assets are a pay as you go basis, and the ability to reallocate these resources to other end users who are willing to pay for them when they are no longer needed by existing end users and small business owners;
2. Having the feeling for the end user and the small business owner that they have unlimited use to IT assets and IT resources;
3. The ability to quickly add or subtract/delete any type of kind of IT resource/IT asset at the most granular level (such as modifying an end user's mailbox through the e-mail interface);
4. With cloud computing resources, the IT assets and the IT resources are not locked into any physical location, so the IT assets and the IT resources can be moved around as needed.

But despite these benefits, it can be quite a complex financial process for a Fortune 100 company to decide if they wish to have their own on-premises IT infrastructure versus provisioning cloud-based resources, as opposed to the small- to medium-sized business, where the decision is so obvious and clear cut.

## Scalability

With regards to the third benefit, that of scalability, this simply refers to the ability for an IT resource or an IT asset to dynamically meet the needs of the end user or small business owner. In most types and kinds of scenarios, this dynamic nature of the IT asset or IT resource can take quite a bit of time to adjust accordingly.

But, when it comes to the cloud computing infrastructure, these particular IT assets and IT resources have to adjust to the need of the end user or the small business owner in just a matter of seconds or less, once their cloud computing infrastructure account has been configured and the desired IT assets and IT resources have been selected and paid for.

This scalability feature of the cloud, and its dynamic nature to meet the needs of the end user and small business owner, is one of the biggest benefits of the cloud computing infrastructure, and one of its strongest selling points to the small- to medium-sized business. Imagine a small- to medium-sized business which possesses an in-house IT infrastructure. The nature and the scope of the business has grown in size, as well as its usefulness and scope. As a consequence, the cost of upgrading the entire IT infrastructure or even just various components of it can be very cost prohibitive for that particular small- to medium-sized business.

But if they had adapted their IT assets and IT resources through a cloud-based infrastructure, they could have brought in the new services needed in just a matter of seconds, and of course, at just a fraction of the cost. And also, it is not just a matter of meeting the specific needs of the end user of the small business owner, the IT resources and the IT assets which are founded and based in the cloud computing infrastructure can also be dynamically met to the needs of the processing fluctuations (of the IT assets and the IT resources) as required.

For example, if an end user requires more disk space, he or she can merely increase it just literally two seconds at the few clicks of the mouse. But, with a traditional, in house IT infrastructure, the entire, physical hard drive would have to be replaced, thus causing the end user an extra expense. This dynamic scalability of the cloud computing infrastructure is a direct function of a concept introduced before, known as proportional costs.

The end user or the small business owner pays for those IT assets or IT resources they specifically use, no more and no less. As a result, the cloud computing infrastructure has adapted the term "as a Service", which will be explained later in this chapter.

## Availability and reliability

One of the last major benefits of the cloud to be reviewed is that of increased availability and reliability. Translated to the language of the small business owner, this simply means that all of the IT resources and IT assets within a specific cloud computing infrastructure is "always going to be on". One may ask at this point is that how can an ISP provide this level of availability to end users and small business owners? Well, think about the actual infrastructure that the ISP resides in.

It is a full blown data center, with dozens of redundant servers, power lines, and even redundant power supply backups, in order to provide a continual supply of power to the data center, so that it's IT resource and IT assets will always be available, and online to end users and small business owners 24 × 7 × 365. But however despite all of these redundancies, and the best efforts which are afforded by the data center, downtimes do occasionally happen.

And in these particular cases, remedies have to be provided to the end user and the small business owner. Another beauty of the cloud computing infrastructure is that even as a particular cloud segment grows and expands the addition of newer IT assets and IT resources and the deletion of older IT assets and IT resources, interruption of the cloud-based services to the end user or small business owner hardly ever happens. It is as if "business as usual is happening".

### SLA agreements

All of this uptime which are guaranteed buy the ISP come to the level of the written contract known as the "Service Level Agreement", or SLA for short. In this contract, the ISP specifies to what levels of uptime it can guarantee to the end users and small business owners. Very often this uptime is defined as the "five 9's", which means that 99.999% of uptime will be made available to the end user or the small business owner. And when this level is not reached, the ISP has to provide remediations to its customer base.

Therefore, it is very important to the customer, whether it is an end user or a small business owner, to read this contract carefully, and before signing it.

## THE CHALLENGES AND RISKS OF CLOUD COMPUTING

Along with these perceived benefits of the cloud, unfortunately, comes its disadvantages well. Remember, the cloud is still a relatively new concept, and the boundary lines it possesses are still not clear. For example, a small business owner's cloud computing infrastructure could actually be shared into another cloud, depending upon the provisioning of the various IT assets and IT resources.

And of course, there is also the trust factor, by placing your entire IT infrastructure into the hands of the ISP, which in this case, could be half way around the world, to people you have never met before, let alone even heard of. And to make matters worse, you are even trusting your credit card information with this third party, which once again you don't even know of. You are even sharing and depending your businesses' confidential and proprietary data, and storing it into the cloud, with other cloud resources.

In this regard, privacy may not be assured. And also, there is always the threat of break ins by hackers who can steal all of your mission critical data. But despite all of this, the large-scale thefts of confidential and proprietary information and data are rare, but it is very important that you, as the end user or the small business owner, understand of these leading, inherent risks, which can be classified as follows:

1. Increased security vulnerabilities;
2. Reduced operational governance control;
3. Limited portability between the cloud processes;
4. Multiregional compliance and legal issues.

## Security risk and challenge

In terms of the first security risk and challenge, which is about the specific security vulnerabilities, the moving of private and confidential business data ultimately becomes the responsibility of the ISP, with which the end user or the small business has entrusted with. True, the contracts and the SLA agreements have been signed, but it still takes quite a bit of faith for the small business owner to transfer all of their corporate information and data to a yet untrusted third party.

This is probably one of the biggest fears that the ISP has to allay. Given the wide expanse of the cloud computing infrastructure, access to IT resources and IT assets cannot be given directly to the end user or the small business owner. Rather, these specific IT assets and IT resources have to be accessed remotely (such as via File Transfer Protocol, or FTP for short), which can pose even greater threats and risks, especially to a hacker who is "listening" in on the other end.

This means that all of the IT resources and IT assets which reside from within the cloud computing infrastructure have to be reached externally, via various access points. The bottom line is that there is no full proof, 100% way to secure those particular access points to the cloud computing infrastructure.

Another major security threat is that of access to confidential consumer data which may be stored in a cloud computing infrastructure, for whatever reason (for example, a retail business owner may be using their cloud infrastructure to back up their customer information and data). How secure this data and information is safe and reliable relies solely upon the security controls and policies implemented by the ISP.

Finally, another grave security threat or risk posed to the cloud computing infrastructure is the intersection of cloud boundaries and the malicious cloud consumers it can create. For example, with shared IT resources and shared IT assets, there are much greater dangers for the theft and damage of mission critical business data stored in the cloud.

The bottom line is that it can be very difficult for an ISP to provide strong security mechanisms when cloud boundaries intersect one another. This cloud computing infrastructure threat is also known as "overlapping trust boundaries", will be discussed more later in this chapter.

## Reduced operational governance

With respect to the second major security threat category, which is that of reduced operational governance control simply refers to the fact that end users and small business owners who use the cloud finally have some control over the IT assets and IT resources which they own or rent. For example, imagine an end user at their place of business or organization. They once had to abide by strict IT security policies. But now, this end user has new founded freedoms to fully control the IT assets and IT assets their business owns or rents in the cloud computing infrastructure.

As a result of this feeling of unfounded power besets the end user, this can translate into real risks for the ISP into how it should control its cloud computing infrastructures. Now, most end users may get a short term rush on this new control they now possess, but these are the end users whose rush can extend into a hacker's mindset and try to gain access to other end user's clouds, and thus to hijack their information and data.

This puts the ISP at graver risk, as well as the lines of communication which exist between the cloud end users and their own cloud computing infrastructures. Two distinct consequences can arise from this: (1) The guarantees as established by the SLA can become null and void, thus causing the ISP significant financial loss and risk; and (2) Bandwidth constraints can occur when the cloud computing infrastructures are located at much further geographic distances from within the platform of the ISP.

One of the best ways, and probably the only way to avoid these kinds of risks and threats is for the ISP to keep a close and vigilant eye on all of the IT assets and IT resources which are being used by all of the cloud users. Any suspicious activity or tampering by would be hackers in the cloud computing infrastructure must be dealt with promptly and deftly, and quickly by the ISP.

## Limited portability

The third major security risk with a cloud computing infrastructure is that of limited portability which are available amongst all of the ISPs. Essentially, what this means is that with all of the ISPs around, there is no common set of standards and best practices amongst them, overall.

Meaning, each ISP can carry its own set of particular technologies at various version levels. While there is nothing inherently wrong with this approach, it can cause major security concerns for the end user and the small business owner, who is wishing to switch to a different ISP. For example, if an end user wishes to transmit and switch over their customer database from one platform to another cloud computing platform at the new ISP, there is no guarantee that this confidential and proprietary information and data will transfer over to the new cloud computing platform cleanly and easily. And in fact, some of this data could even be hijacked by a third party, such as a hacker.

A lot of this reason has to do with the fact many cloud computing architectures and platforms are to a certain degree, proprietary from one ISP to another. Let us look back at out previous example once again.

Suppose that the small business owner has a customer database which is custom made and is dependent upon other various technology platforms. The new ISP may flount serious claims that this customer database will transfer smoothly over to their own proprietary cloud architecture. But, the truth of the matter is, what if this customer database does not transfer cleanly from the existing cloud computing infrastructure to the new cloud computing infrastructure?

True, the small business owner could be looked down upon for not doing their part of due diligence to look at the technological compatibilities for this particular kind of database, but the real accountability will rest with the new ISP. What lacks between these ISPs is a set of best practices and standards which allow for the ease of transfer of information and data for the end user from one cloud computing infrastructure to another ISP.

But at least for now, the ISPs are starting to realize this big gap, and in lieu of it, are now offering cloud computing technologies which are standard amongst the industry. Having this level of standardization is especially important for the biometric templates, especially when they are stored in a cloud computing environment.

With regards to the biometric templates, any loss of any magnitude of any type or kind from the transfer of one cloud computing infrastructure to another cloud computing infrastructure will have far reaching and more reaching legal ramifications in both the courts of law and judicial processes.

## Compliance and legal issues

The fourth major security vulnerability posed to a cloud computing infrastructure is that of the multiregional compliance and legal issues which can arise. What does this exactly mean? It refers to an aspect which was discussed in the earlier part of this chapter. And that is, the cloud computing infrastructure, or for any matter related to it, are located in different geographic regions around the world.

For example, although a small business or organization might be based in the United States, the owner of that particular business or organization can purchase a cloud computing infrastructure from a geographic halfway around the world, such as Germany or Russia. The advantages of this course are the very cheap prices which can be offered to the small business owner, as it has been discussed earlier in this chapter. But really, to the small business owner, it makes no difference where the actual, physical servers are located from within the cloud computing infrastructure.

While most of the time this is not the case, but the information and data which is housed in a cloud computing infrastructure in a different country could very well be suspect and prone to the laws of that particular country as to where that cloud computing infrastructure actually resides. To this effect, these foreign governments could very well improve their own data privacy and storage policies.

Another strong security concern is that of the accessibility and disclosure of the information and data which resides in the cloud computing infrastructure. For example, if a small business owner in Europe owns a cloud computing infrastructure here in the United States, their information and data could be prone to inspection by the United States Government based upon the guises of the US Patriot Act. This is one of the weakest areas for putting a biometric system into a cloud computing infrastructure. This is so because the storage of all of the biometric templates on a cloud computing infrastructure is a very sensitive topic by in of itself, and if this cloud-based biometric database does come under heavy scrutinization by a government agency, cries of Civil Liberties and Privacy Rights violations will abound, thus defeating any purpose of a biometrics cloud-based infrastructure.

But of course, this also brings up the issue of whether a biometrics cloud-based infrastructure should be stored in a public cloud versus a private cloud. This will explained in much more detail later in this chapter.

## THE FUNCTIONS AND CHARACTERISTICS OF CLOUD COMPUTING

So far in recap of this chapter, we have reviewed the definition of what cloud computing is, the major components which make it up, as well as the benefits and the major risks in terms of security which are posed to a cloud computing infrastructure. Now, in this part of the chapter, we turn our attention over to the characteristics which separate the cloud computing infrastructures from the actual IT infrastructures which exist today.

As we have eluded to throughout this chapter, the cloud, to many people, is a very nebulous concept. Nebulous in this case means very murky, no clear boundaries, nor any clear definitions. True, in many aspects, this is what the cloud computing infrastructure is all about. It comes from out of somewhere, in some part of the United States, or for that matter, in any part of the world, even in the most remote region imaginable.

All one has to have is a computer, an Internet connection, and literally from thousands and thousands of miles away, the access to an ISP. With all this, the end user or the small business owner, at the click of a mouse, has to do is just merely select the IT assets and the IT resources they need or desire.

In a way, it can be a scary sort of thing, after all, we are entrusting to store our valuable and confidential data to people at the ISP to whom the end user or the small business owner has never even met or even heard of. But, it takes a giant leap of faith to have this kind of level and caliber of trust. But despite this degree of uncertainty, the cloud computing infrastructure does possess a number of quantifiable, key characteristics, which help it to define and separate itself from other IT Infrastructure regimes.

These characteristics include:

1. On Demand Usage;
2. Ubiquitous Access;
3. Multitenancy and Resource Pooling;
4. Elasticity;
5. Measured Usage;
6. Resiliency.

## On demand usage

The first cloud computing characteristic, which is On-Demand Usage, refers to the fact that if an end user or small business owner provisions certain cloud computing services onto their account, that this particular IT resource or IT asset will be made immediately for use after it has provisioned. With the characteristic of on demand usage, the end user can thus request future IT assets and IT resources on literally a automatic basis, without any human intervention at the ISP what so ever.

With this characteristic of an on demand usage for the cloud computing infrastructure, the end user or the small business owner is literally free to turn them on or off the IT resources and IT assets when they want to or need to at complete will, and pay for only what is used.

## Ubiquitous access

The second characteristic of the cloud computing infrastructure is known as what is Ubiquitous Access. This simply means that the IT resources and the IT assets are widely accessible from anywhere on earth, and can be reached via any mobile device via any network protocol. This can best be illustrated by a small business owner accessing their cloud-based IT resources and IT assets through the other side of the world, with their mobile device, via a Wireless Network Protocol (WAP).

## Resource pooling

The third major characteristic of a cloud computing infrastructure is that of what is known as Multitenancy and Resource Pooling. The basic definition of multitenancy is "a characteristic of a software program that enables an instance of the program to serve different consumers (tenants) whereby each is isolated from the other" (Source: *Cloud Computing: Concepts, Technology & Architecture*, Erl, T., Arcitura Education, 2013, p. 59). Take for example an application we have eluded to earlier in this chapter before, which is that of Microsoft SQL Server.

There may be just one instance of it running on the physical server, but this particular instance allows to serve the different and multiple end users and small business owners. This example demonstrates multitenancy, and required the principles of Virtualization to make it all happen. Although one instance of SQL Server is on the physical server and shared with others, the end user or the small business owner get to feel that this is their very own software through their cloud computing infrastructure control panel.

In fact, it is this multitenancy model which allows for the cloud-based IT resources and IT assets to be dynamically assigned over and over again, based upon the demands and the needs of the end user and the small business owner. The multitenancy model as a result, has given rise to another concept called Resource Pooling. This allows for "cloud providers to pool large scale IT resources to sere multiple cloud consumers. Different physical and virtual IT resources are dynamically assigned and reassigned according to cloud consumer demand, typically followed by execution through statistical multiplexing" (Source: *Cloud Computing: Concepts, Technology & Architecture*, Erl, T., Arcitura Education, 2013, p. 59).

These multitenancy and resource pooling concepts allows the end users and the small business owners to use the same IT assets and IT resources while each end user and small business owner remains unaware that the same IT resources and IT assets are being used by others as well.

## Elasticity

The third major characteristic of a cloud computing infrastructure is that of Elasticity. When one thinks of Elasticity, very often, the image of a rubberband is conjured up. This is perfectly analogous to what a cloud computing infrastructure should be like. It should be literally be able to flow smoothly, and have the ability to scale to the needs and wants of the end user or the small business owner.

For example, if an ISP expects to keep its customer base and attract newer technologies into its cloud computing infrastructure, it must be flexible enough to keep up with the changing market conditions and demands which are placed upon it. An ISP which is not flexible and are rigid in their ways of technology will be sure to lose out in terms of keeping its competitive advantage.

It should be noted that ISPs with vast resources for a cloud computing infrastructure offers the most flexibility and elasticity. The elasticity of a cloud computing infrastructure can also be best measured by how quickly an end user or a small business owner can turn off and turn on their new cloud-based services.

## Measured usage

The fourth major characteristic of a cloud computing infrastructure is that of Measured Usage. As its name implies, this feature simply takes into account how much usage the end user or the small business owner is taking out the IT assets and IT resources that they have selected to use. Based upon that particular usage, then the end user and small business owner is appropriately charged for the cloud computing infrastructure resources utilized.

Normally, this is flat fee on a monthly basis, but if usage of the IT resources and IT assets go beyond the expected levels, of course, the charges will be much higher. But on the converse, if the usage of the IT assets and the IT resources fall below the expected usage level, the end user or the small business owner still has to pay the flat monthly fee.

In this regard, a cloud computing infrastructure possesses the "use it or lose it" feature, unless the specific services have been cancelled. But the characteristic of Measured Usage does not necessarily mean how much to charge, it also refers to the overall usage of an IT asset or an IT resource being utilized, and those not being used. That way, future demand can be predicted, and the cloud computing resources can be provisioned accordingly.

## Resiliency

Finally, the fifth and final characteristic of a cloud computing infrastructure is that of Resiliency. This simply means that "IT resources can be pre-configured so that if one becomes deficient, processing is automatically handed over to another redundant implementation" (Source: *Cloud Computing: Concepts, Technology & Architecture*, Erl, T., Arcitura Education, 2013, p. 61). In the scheme of cloud computing infrastructure, it also means that there are redundant IT resources and IT assets available, so should one set fail, the end user and the small business owner will suffer 0% downtime.

In other words, smooth and seamless operations are always present in a cloud computing infrastructure.

## CLOUD COMPUTING DELIVERY MODELS

Now, that we have reviewed some of the major characteristics of a cloud computing infrastructure, it is now time to understand how such a cloud computing infrastructure can be brought into the hands of the end user, or the small business owner.

Although the cloud structure does involve in being in a nebulous state, the delivery method is not so. There are many ways in which the cloud can be brought to the end user or the small business owner, there are three common types of platforms.

These platforms include:

1. Infrastructure as a Service (IaaS);
2. Platform as a Service (PaaS);
3. Software as a Service (SaaS).

Each of these platforms has their own unique role in delivering the cloud computing infrastructure, and they will be discussed in length in this section. With the first cloud computing infrastructure deployment model, which is the IaaS, represents the overall cloud computing environment infrastructure. In other words, IT resources and IT assets "can be accessed and managed via cloud based interfaces and tools" (Source: *Cloud Computing: Concepts, Technology & Architecture*, Erl, T., Arcitura Education, 2013, p. 64)

As its name implies, the IaaS provides the framework or the foundation from which all of the IT assets and IT resources can be leveraged towards the small business owner and the end user. This infrastructure, in particular, includes the hardware, network connectivity, all of the software applications (which includes, for example, all of the VoIP applications, e-mail applications, database applications, and software development applications) as well as the other "raw" tools which helps to comprise the IaaS Infrastructure.

It should be noted that most IaaS IT assets and IT resources are "virtualized", and bundled in such a package that they can be leveraged through the cloud to the end user or the small business owner. Thus, these virtualized IT assets and IT resources can have the freedom of scalability, customization, and demand availability, which are all of course, very crucial components that a cloud computing infrastructure needs to have.

By possessing an IaaS Infrastructure, the end user or the small business owner can have total control and responsibility over their particular cloud computing infrastructure. For example, once an end user or a small business owner signs up for a cloud computing infrastructure account, they are often given access to a control panel from which they can establish the settings, the permissions, and even install and uninstall particular cloud computing resources.

Every ISP gives this tool to all of their customers. After all, this is the only direct way for the end user or the small business owner can access all of their subscribed to IT assets and IT resources. It should be noted that with the IaaS platform, the end user or the small business owner assumes full administrative control over their cloud-based IT assets and IT resources.

For example, when an end user or a small business owner first signs up for their cloud computing infrastructure account, the IT assets and the IT resources that they will be subscribing to are what is known as "fresh virtual instances". Let us demonstrate with an example. Suppose that an ISP has literally hundreds of physical servers. These servers contain all of the IT software-based IT resources and IT assets.

Once the end user or small business owner provisions their own account, the hard drive from which the software-based IT resources and IT assets will be distributed from will be divided into its own partition on the physical server. Thus, this will give the end user or small business owner the look, the feel, and the total control over their own, unique cloud computing server.

On the converse, if an end user or small business owner needs a gargantuan account of IT assets and IT resources, they can lease out an entire cloud-based server, which in this scenario is known as a "Virtual Server", in the IaaS Platform.

## PaaS

The second deployment model for the cloud computing infrastructure is known as PaaS for short. Specifically, it can be defined as "a predefined 'ready to use' environment typically comprised of already deployed and configured IT resources" (Source: *Cloud Computing: Concepts, Technology & Architecture*, Erl, T., Arcitura Education, 2013, p. 65). The prime differentiation between the PaaS Platform and the IaaS Platform is that the latter consists of the raw cloud computing platform.

In other words, as it was reviewed previously, the IaaS contains the basic materials needed for the foundation of a cloud computing infrastructure. Imagine if you will, the IaaS Platform which serves as the foundation for the cloud computing infrastructure. The PaaS Platform fills up this foundation with the much needed IT resources and IT assets in order to fulfill the needs of the end user or the small business owner.

As it can be described, the PaaS consists of a set of prepackaged IT products and IT tools to help support the business needs of the end user or the small business owner. There are many reasons why a consumer should choose the PaaS, but the following are the most typical reasons:

1. Having the sense of scalability;
2. The client can use the literally ready to use environment, and specifically modify it to their own needs and wants;
3. If the end user or small business owner feels confident enough, the can even their own services, through the use of the PaaS, to other cloud consumers.

The PaaS permanently alleviates on part of the end user or the small business owner the need to fill up and have the overall responsibility of administering the IaaS. Rather, this option is up to the ISP to provide to its consumer base. So, as you can see, it is the PaaS which contains all of the needed IT resources and IT assets that the end user or small business owner can choose from. So, to summarize thus far, the IaaS is a service which provides the foundation for a cloud computing infrastructure, and the PaaS makes the IT assets and IT resources available to the end user or the small business owner wither as a deployed cloud package or a single service. This latter is known as "Software as a Service", and will be discussed next.

## SaaS

The next cloud computing infrastructure to be looked at is the SaaS, as eluded to before. As we had previously discussed earlier in this section, the IaaS provides the foundation for the cloud computing infrastructure, the PaaS fills up this foundation, with all of the IT assets and IT resources that would fulfill the needs of all types and kinds of end users and small business owners. Now granted, these end users and small business owners will not need to use all of the IT resources and IT assets which are available in the PaaS, so thus, this is where the SaaS Platform comes into play.

Specifically, the SaaS model in the cloud computing infrastructure can be defined as "a software program positioned as a shared cloud service and made available as a 'product' or general utility which represents the typical profile of an SaaS offering" (Source: *Cloud Computing: Concepts, Technology & Architecture*, Erl, T., Arcitura Education, 2013, p. 66). So, as you can see from the definition, the SaaS component of the cloud computing infrastructure can much be viewed as a "marketplace" in which the end user or the small business owner can literally cherry pick the IT assets and the IT resources they need or desire.

At the SaaS level, the small business owner or the end user can pick all of the software packages or bundles they require in order to keep their business running smoothly, and to maintain a competitive advantage. For example, one such SaaS offering which is popular is the hosted e-commerce store. With this, a small business owner can attract many customers. But despite all of these advantages of the SaaS, it does have serious limitation.

The end user or the small business owner has very little control over their IT resources and IT assets that they have selected from the SaaS Platform. This control is restricted to administrative control, after the IT assets and the IT resources have been selected and paid for. There is very little that can be done in the way of administration of those IT assets and IT resources which reside in the SaaS.

In summary, we have looked at three major cloud delivery models:

1. The IaaS: This lays the foundation, or the nuts and bolts of the cloud computing infrastructure;
2. The PaaS: This model provides all of the IT resources and IT assets which an ISP can provide to its consumers. All of these resources and assets include everything from the virtual server to the hard drive partitions which provide the end user or the small business owner with the look and feel of their own cloud-based server, as well as the software applications which range from the content management systems to the e-commerce platforms;
3. The SaaS: As reviewed, this model provides the software applications (also known as the IT assets or IT resources) via an a la carte method to the end user or the small business owner. Under this regime, the customer can pick and choose the software applications which are needed or desired, and within seconds, it can be provisioned at the click of the mouse after it has been paid for.

## CLOUD COMPUTING DEPLOYMENT MODELS

Now that we have reviewed how a cloud computing infrastructure can fit into a certain platform, we now turn our attention to how the cloud computing infrastructure can be handed over to the small business owner or the end user.

In other words, the cloud computing infrastructure is now defined via three different infrastructure models, but now, it has to reach the power of the end user or the small business owner via the network to their computer, laptop, netbook, or even smartphone. The cloud computing infrastructure is brought to the end user or small business owner via four different kinds of deployment models which are:

1. The public cloud;
2. The community cloud;
3. The private cloud;
4. The hybrid cloud.

## Public cloud

We will now discuss each of these cloud computing infrastructure deployment models in much greater detail. In terms of the cloud computing infrastructure, the public cloud can be specifically defined as "a publicly accessible cloud environment owned by a third party cloud provider" (Source: *Cloud Computing: Concepts, Technology & Architecture*, Erl, T., Arcitura Education, 2013, p. 73).

As the name implies, this type of cloud computing infrastructure is available via any type or kind of ISP. The IT assets and the IT resources as described throughout this chapter can be provisioned amongst all three delivery models just described (IaaS, PaaS, SaaS), and are also available to the end users and the small business owners at almost very low cost or price, and the IT assets and the IT resources are very often "commercialized" via other methods such as advertisements on the Internet, or in print, such as magazines and newspapers.

With the public cloud deployment model, the ISP has the primary responsibility of this kind of deployment, the acquisition, and the procurement of the IT assets and IT resources which belong in the PaaS Platform, it's ongoing maintenance (such as licensing fees and any other necessary types or kinds of software and hardware upgrades).

It should be noted that all (about 98% or so) cloud computing infrastructure offerings are offered to the small business owner and the end user via this type of deployment model, so that all types and kinds of IT resources and IT assets can be made available plentifully and at low cost to all entities involved.

## Community cloud

With the second type of deployment, which is that of the community cloud, is also very much related to that of the public cloud. But, access to this particular type of cloud computing infrastructure is sheerly limited to a specific grouping of end users and small business owners. It should be noted that at this point the community cloud could be jointly owned by the members of the cloud community or even an ISP could own this community cloud.

The members of the community cloud typically share the responsibility for defining the rights and the responsibilities of the community cloud. However, it should be noted also that membership into the cloud community is not a status, and that membership into it does guarantee access to the IT resources and the IT assets which reside in the cloud computing infrastructure.

## Private cloud

The third type of cloud computing deployment model is that of the private cloud. The private cloud computing infrastructure is literally owned by a single entity. With this type of cloud computing infrastructure deployment model, this allows for a small business owner or any other kind of organization the total control over the IT assets and the IT resources from within the different locations and departments at the place of business or organization.

Typically, under a private cloud computing infrastructure deployment model, the administration of it can take place internally, thus giving the private cloud owner even more sense of ownership and control. From within the private cloud computing infrastructure deployment model, it is typically both the place of business or organization which is both the consumer and the cloud owner.

To help the place of business define and differentiate these roles between consumer and cloud owner, the following guidelines are established:

1. A separate department from within the place of business or organization assumes the cloud provisioning duties to be performed. These groups of people become the private cloud owners;
2. All of the departments at the place of business requiring access to the IT resources and IT assets from within the Private cloud computing infrastructure become the cloud consumers.

## Hybrid cloud

The next type of cloud computing infrastructure deployment model we examine next is the known as the "Hybrid Cloud". As its name implies, the hybrid cloud consists of a combination of the various cloud deployment models just discussed, which include the community cloud, the public cloud, and the private cloud. The hybrid cloud does not have to consist of all these cloud types, but it must contain at least a combination of two types of cloud deployment models, and from that, any other cloud deployment model can work just as well, with the different deployments of IT assets and IT resources.

For example, a small business owner can choose to protect their confidential and private data (such as financial information and customer data) from within the private cloud infrastructure. Now, if this small business grows in size over time to a multinational corporation, the business owners can then develop the private cloud into a hybrid cloud computing infrastructure and put it into a community cloud, and create more if necessary.

So as you can see, trying to implement a hybrid cloud deployment model can become very complex and much more of a challenge to administer and maintain because of the differing cloud environments which are presently available. Because of this, an alternative exists to the hybrid cloud – this is called the virtual private cloud.

Under this cloud regime, the end user or the small business owner is given a self-contained cloud environment which is dedicated solely to their uses, and also are given full administrative rights to do what the cloud end users or the small business owner wishes to do their in their own, particular virtual private cloud environment.

## THE SECURITY THREATS POSED TO CLOUD COMPUTING

Now that we have covered in great detail and depth what the cloud computing delivery and deployment models are like, we now turn our attention to the weaknesses that are posed to the cloud. More specifically, these can be categorized as "Threat Agents". But before we divulge deeper into what a threat agent is specifically for a cloud computing infrastructure, it is first very important to review some very basic terms and concepts as it relates to cloud computing threat agents.

Specifically, these terms and concepts include the following:

1. Confidentiality;
2. Integrity;
3. Authenticity;
4. Availability;
5. Threat;
6. Vulnerability;
7. Risk;
8. Security controls;
9. Security mechanisms;
10. Security policies.

These above concepts and definitions form the basics for understanding what threat agents are to a cloud computing infrastructure. The above definitions and concepts will be reviewed and examined in more detail further, but it is also equally important to review the threat agents posed to an overall cloud computing infrastructure.

These threat agents include:

1. Anonymous attacker;
2. Malicious service agent;
3. Trusted service attacker;
4. Malicious insider.

Apart from the above cloud computing threat agents, there are also more specific security threats to a cloud computing infrastructure, and they are as follows:

1. Traffic eavesdropping;
2. Malicious intermediary;

3. Denial of service;
4. Insufficient authorization;
5. Virtualization attack;
6. Overlapping threat boundaries;
7. Flawed implementations;
8. Security policy disparity;
9. Contracts;
10. Risk management.

## Confidentiality

All of these concepts and terminologies will now be reviewed in much greater detail. With respect to the very basic terms and fundamentals, the first which is of importance is known as "Confidentiality". In simple terms, this means that the IT assets and the IT resources are only made available to authorized users. In this case, it means that the paying subscribers such as the end users and the small business owners only have access to those IT assets and IT resources that they have paid for, especially when those resources and assets are in transit across the network medium, and that after access remains very strict when it comes to remote logins.

## Integrity

With regards to Integrity, this means that the IT assets and the IT resources which reside at the ISP are not altered or changed in any way or form. But more importantly, it is the end user's and the small business owner's data and information which is being processed at the virtual servers located at the ISP which need strong assurance this confidential and private information is not altered in any way, or in any malicious form. In other words, "integrity can extend to how data is stored, processed, and retrieved by cloud services and cloud based IT resources" (Source: *Cloud Computing: Concepts, Technology & Architecture*, Erl, T., Arcitura Education, 2013, p. 119).

## Authenticity

With regards to Authenticity, this proves that all of the IT resources and IT assets are legitimate, and that all of the of information and data processed at the virtual server at the ISP and all of the output which comes from it is genuine, and that all this has been provided to the end user or the small business owner from a legitimate cloud computing infrastructure rovider.

## Availability

In terms of Availability, this means that the IT assets and the IT resources will always be accessible and be rendered usable whenever and wherever the end user or small business owner needs to access them at a second's notice. This uptime is very often spelled out in the SLA that the end user or the small business owner must subscribe to.

With regards to threats, it is a threat agent of sorts which can fatally challenge the defenses afforded to the cloud computing infrastructure. It should be noted that the threat just defined if carried out with a malicious intent is known as a "cyberattack".

## Vulnerability

In terms of the next important security foundation, vulnerability is a weakness in the cloud computing infrastructure which can be easily overcome by a cyberattack, as just previously described. A vulnerability in the cloud computing infrastructure can be caused by a whole host of reasons, which can include both software and hardware misconfigurations, very weak security policies, administrative or user errors, hardware and software deficiencies, and an overall poor security architecture for the cloud computing infrastructure.

## Security risk

In terms of security risk to the cloud computing infrastructure, this represents the statistical probability of a loss occurring to the cloud computing infrastructure. Cloud computing risk is typically measured according to the threat level, or as a whole when compared to the total number of possible or known threat agents. There are two known metrics used to determine the statistical probability of risk:

1. The statistical probability of a threat occurring to exploit the IT assets and the IT resources in the cloud computing infrastructure;
2. The chances of the expectations of a magnitude of a loss occurring from the IT resource or the IT asset within the cloud computing infrastructure.

## Security controls

With regards to security controls, there are countermeasures which are employed to protect the cloud computing infrastructure and the IT assets and the IT resources which reside within it. The details on how to implement and use such security controls from within a cloud computing infrastructure is often spelled out in the security policy.

## Security mechanisms

In terms of security mechanisms, there are countermeasures and protective mechanisms used to provide the general framework which protect the IT assets and the IT resources in the cloud computing infrastructure.

## Security policies

Finally, the security policies associated with a cloud computing infrastructure are the security rules and regulations which govern its safety. It is the cloud computing security policy which further the types and kinds of rules and regulations which are needed to provide maximum security to the cloud computing infrastructure.

Now that we have looked at some of the very important terms and concepts which underpin a cloud computing infrastructure, as eluded to before, we now turn our attention over to the points of vulnerability. These are those specific entities that can carry out malicious activity against a cloud computing infrastructure, and provide great harm to the end user or the small business owner.

The first type of these types of entities are known now specifically as the "Threat Agent". As the term implies, this of entity can carry out a direct attack against a cloud computing infrastructure. What is even more dangerous about this type of attacker is that this entity can originate wither internally within the place of business or organization, or just even outside of it.

## Anonymous attacker

The first formal type of threat agent is known as the "Anonymous Attacker". This type of attacker is a malicious third party without any specific cloud-resource-based permissions. This type of attacker exists as a software program, and can launch specific attacks to the virtual servers via the public network. These types of attack entities have limited information or knowledge about the security defenses of the ISP, so the best way they can get into a cloud computing infrastructure is by hacking into and stealing user accounts.

## Malicious service agent

The second type of formal threat agent is known as the "Malicious Service Agent". This type of entity typically exists from within the cloud computing infrastructure, and has the capability to intercept and forward network traffic. Again, this type of threat agent usually exists as a software program.

## Trusted attacker

The third type of formal threat agent is specifically known as the "Trusted Attacker". As the name implies, the Trusted Attacker is already a consumer of a particular cloud computing infrastructure, and therefore, they have advanced knowledge (to some degree) of its inner workings. This type of attacker usually launches their attacks from within a cloud computing infrastructure, very often using their own login information from within a cloud computing infrastructure, very often using their own login information and access credentials. The main threat of a trusted attacker is that given their knowledge of the cloud computing infrastructure, they can attack just about every type and kind of virtual server from within the cloud computing environment.

## Malicious insider

The final type of threat agent is known as the "Malicious Insider". Typically, these are contracted employees or ex employees where login and access information into the particular cloud computing infrastructure has not been deleted as of yet. Given this short window of timeframe until their access information is totally deleted, these type of attack entities can still log into their former cloud computing infrastructures and still cause great moments of harm and damage.

## Traffic eavesdropping

Now, that we have seen what some of the major threat agents are to a cloud computing infrastructure, we now turn our attention to some of the major, actual threats which are posed to a cloud computing infrastructure. The first type of threat is known as "Traffic Eavesdropping". This simply happens when the end user's or small business owner's information and data which is still being transferred to their cloud computing infrastructure is covertly intercepted by a third party (such as a malicious hacker) for the illegal purposes of misusing and abusing that particular information and data.

The pure purpose of this kind of attack is to breach the relationship between the end user/small business owner and the ISP. The traffic eavesdropping security threat very often goes unnoticed for very long periods of time.

## Malicious intermediary

The second type of threat to a cloud computing infrastructure is known as a "malicious intermediary". In this type of attack scenario, the end user or the small business owner's information and data is intercepted by a malicious third party, and is intentionally altered, thereby compromising the integrity of that particular information and data, with the end resultant being totally spoofing the consumer of the integrity of that information and data.

## Denial of service

The third type of threat to a cloud computing infrastructure is that of "denial of service". The primary objective of this type and kind of attack is to constantly overload and bombard both the physical and virtual servers up to the point they cannot operate anymore, and literally shut down, thus depriving the end user and small business owner of their IT assets and IT resources.

Here is how a denial of service attack could conceivably work:

1. The physical servers and the virtual servers are totally bombarded with malformed data packets;
2. Subsequently, the network traffic is greatly reduced and responsiveness is totally choked off between the ISP and the end user/small business owner connectivity to their IT assets and IT resources;
3. Next, the memory of both the physical and the virtual servers is totally bogged down by processing the malformed data packets and then shuts down, thus cutting off all types and kinds of access.

## Insufficient authorization

The fourth type of attack to a cloud computing infrastructure is that of "insufficient authorization". This is when an attacker is granted access by the sheer mistake made by the ISP, thereby giving them all access to all of the physical and virtual servers, and IT assets and IT resources. Another type of attack of this kind is known as "weak authentication". This occurs when there is very low entropy, or weak passwords which are being used by the end user or the small business owner to access their specific IT assets and IT resources.

## Authorization attack

The fifth type of attack to a cloud computing infrastructure is that of the "virtualization attack". Because with the nature of the cloud computing infrastructure, the end user and the small business owner have to a certain degree, full administrative privileges. Because of this inherent risk, the end user or the small business owner can become a cloud computing infrastructure threat in of themselves.

More specifically, the virtualization attack takes pure, 100% advantage of the weaknesses of the confidentiality, integrity, and availability platforms of the virtual servers which reside within the cloud computing infrastructure.

### Overlapping trust boundaries

The fifth particular type of attack to a cloud computing infrastructure is what is known as "overlapping trust boundaries". One must remember that with a cloud computing infrastructure, all of the IT assets and IT resources are shared amongst one another, even though the end user or the small business owner gets the look and feel as if it is their very own.

So given this shared resource nature, a small business owner or end user who has an extensive IT background and a very strong malicious intent can easily take advantage of the vulnerability of shared resources, launch an attack, and totally corrupt the information, data, and even bring down the virtual servers of other end users and small business owners. In other words, the goal here is to literally target the overlapping trust boundaries between the various cloud computing infrastructures which exist at the ISP.

## IMPORTANT MECHANISMS OF CLOUD COMPUTING

So far in this chapter, we have examined a lot if concepts and theories which deal with a cloud computing infrastructure. We have looked at everything from the various cloud models to the deployment mechanisms, as well as some of the major threats and risks which are posed to a cloud computing infrastructure. And so far in our examination of a cloud computing infrastructure, we have looked at it in terms of a measured service. In other words, what it can do for the specific needs of the end user or the small business owner.

But, the cloud itself has inner workings of its own which enable to function the way it does. These components of the cloud computing infrastructure can be referred to as specialized cloud mechanisms. And in this part of the chapter, we review six major ones (there are of course many others, but these five are most important) and they include the following:

1. The load balancer;
2. Pay per use monitor;
3. The audit monitor;
4. The failover system;
5. The hypervisor;
6. Resource clustering.

## Load balancers

With the load balancers mechanisms in the cloud computing infrastructure, rather than place the strain on just one IT asset or just one IT resource, the workload is placed upon two or more IT resources to help increase the performance and capacity. The load balancers can perform the following workload distributions:

1. Asymmetric distribution: The larger workloads are given to those IT assets and IT resources which have more sophisticated processing power capabilities;
2. Workload prioritization: The cloud computing infrastructure workloads are scheduled and distributed according to the priority level;
3. Content-aware distribution: The requests set forth by the end user and the small business owner are sent over to the various IT asset and IT resources in the cloud computing infrastructure.

It should be noted load balancer is pre-configured with performance rules for the optimization of the client-based IT assets and IT resources and to avoid overload in the cloud computing infrastructure. These rules can exist as:

1. A multi-layered switch;
2. A dedicated hardware appliance;
3. Dedicated software;
4. A service agent.

It should be noted that the load balancer is usually located in the end user/ small business owner's IT resources and IT assets which are generating the workload and performing that actual workload.

## Pay per use monitor

With regards to the second mechanism, the pay per use monitor, "is a mechanism which measures cloud based IT resource usage in accordance with predefined pricing parameters and generates usage logs for free calculations and billing purposes" (Source: *Cloud Computing: Concepts, Technology & Architecture*, Erl, T., Arcitura Education, 2013, p. 184). This monitor can include:

1. The data volume;
2. The bandwidth consumption.

It should be noted that the pay per use mechanism is monitored by the billing department by the ISP, in order to transmit the proper fees to the end user or the small business owner.

## Audit monitor

In terms of the third monitor, which is the audit monitor, this type of mechanism collects and analyzes various types and kinds of data and network traffic and activity within the cloud computing infrastructure, which are primarily dictated by the regulations set forth in the SLA agreement.

The fourth kind of cloud computing infrastructure mechanism is that of the cloud-based failover system. With this type of system, it is used to increase the availability of the IT resources and the IT assets which are made available to the end user and the small business owner. The basic premise is to provide large amounts of redundancy in case of a failover, so that all of the IT assets and the IT resources will always be available 24 × 7 × 365.

It is purely designed to rollover to another IT asset or IT resource in case one fails. In other words, the failover system provides redundancy so that all cloud-based assets and resources will always be available to the end user and the small business owner. It should be noted that failover systems are used primarily in mission critical programs and applications. A redundant cloud computing infrastructure mechanism can span thousands of miles across many countries and in many divisions of the ISP in order to provide the 24 × 7 × 365 redundancy for the IT assets and the IT resources. The failover or the redundancy systems for both the IT assets and the IT resources come into two basic types and configurations:

1. Active–active
2. Active–passive

With the active–active failover or redundant system, many instances of the same IT asset and IT resources actively synchronously together. Then one type of failover occurs with one or more of the IT assets and IT resources, then those failed IT assets and IT resources are removed from what is known as a load balancer scheduler. After the failed IT resources and IT assets are removed, then the remaining IT assets and IT resources still continue in the operational mode.

With the active–passive failover or redundant system, not all of the IT assets and IT resources remain active. Only a portion of them remain active. So, in this type of configuration, the inactive IT assets become available or active when any of the other IT resources and IT assets fail and the processing workload is this redirected to those IT assets and IT resources which become active and functionable.

It should be noted that in both types of failover or redundant systems, switching over to a new IT asset or IT resource is not done until at least a new IT asset or IT resource thoroughly dies out. However, there are cloud computing infrastructure failover or redundant systems which can detect failover conditions before total failure occurs and literally shuts down that IT asset or IT resource before it fails.

## Hypervisor

The fifth very important type of cloud computing infrastructure mechanism is that of the hypervisor. This is a very important tool in cloud computing, as the hypervisor is utilized to create or generate virtual servers as offset instances of the physical server system. It should be noted that the Hypervisor can only be used on one type of physical server, and virtual images can be created of only that one particular server. Also, any IT asset or IT resources which are created from that Hypervisor as a result of the virtual server generation, are tied or associated only to that particular physical sever.

## Resource clustering

The sixth cloud computing infrastructure mechanism we will now review is that of resource clustering. Typically, all cloud-based IT assets and IT resources are set very far apart, geographically speaking. Despite this physical distance, these servers, in the virtual sense, can logically be combined into groups known as "resource clusters" in order to greatly improve the availability of IT assets and IT resources to the end user and small business owner.

In this regard, it is the resource cluster mechanism which groups multiple IT assets and IT resources into a single cloud computing infrastructure instance. This clustering effect increases cloud computing capacity, load balancing, and the overall availability of the IT assets and the IT resources to the end user and the small business owner.

There are three types of cloud-based cluster types:

1. Server cluster;
2. Database cluster;
3. Large dataset cluster.

## Server cluster

With the server cluster scheme, both the physical and virtual servers are all clustered together in one large-scale effort to boost IT resource and IT asset performance and availability. In terms of database clustering in the cloud computing infrastructure, this is designed and created to help improve the data availability which is owned by the end user or the small business owner. The database cluster consists of a synchronization feature which helps to ensure the consistency of the data information which can be stored at various and different storage devices which are contained in the database cluster.

With a large dataset cluster, the data partitioning and the distribution is set forth so that the end user's and small business owner's datasets can be efficiently partitioned without compromising the structure or the integrity of the datasets.

Now, there are two basic types of resource clusters:

1. Load balanced cluster;
2. HA cluster.

In the event of a multiple failure event at the ISP, the load balanced cluster possesses redundant implementations of all of the clustered IT assets and IT resources. With the HA system, a failure system mechanism is quite often used, and this monitors failure conditions and will automatically redirect any cloud computing infrastructure workload from the failed IT assets and failed IT resources.

## Cloud computing cost metrics and service quality mechanisms

Now that we have looked into what makes a cloud computing infrastructure work from the inside as well as the outside, we now turn our attention to another very important aspect of the cloud computing infrastructure – which are the metrics which are associated with them. Why are metrics so important? Well, it is to give a sense of protection to both the end user and the small business owner, knowing that there is something they can gauge, or compare with other ISPs who are also offering the levels of IT assets and IT resources.

In other words, the end user and the small business owner, given these metrics, can now quantitatively benchmark to see what will work best for them. In this chapter of the book, we look at both Cost Metrics and Service Quality Metrics. But first, we look at Cost Metrics in much more detail, which include the following:

1. Network usage;
2. Server usage;
3. Cloud storage device usage;
4. Other cost management considerations.

## Network usage

In terms of network usage, it is defined as the "amount of data transferred over a network connection, network usage is typically calculated using separately measured inbound network usage traffic and outbound network usage traffic metrics in relation to cloud services or other IT resources" (Source: *Cloud Computing: Concepts, Technology & Architecture*, Erl, T., Arcitura Education, 2013, p. 387).

As it is described in the definition, network usage consists of the:

1. Inbound network usage metric:
   a. It measures the amount inbound network traffic;
   b. This metric is measured in bytes.
2. Outbound network usage metric:
   a. This measures the amount of outbound network traffic;
   b. This metric is measured in bytes.
3. Intra-cloud WAN usage metric:
   a. This metric measures the amount of network traffic of the various IT assets and IT resources which are located in geographic diverse segments;
   b. This metric is measured in bytes.

It should be noted that the network usage metric are ascertained by the following components:

1. Static IP address usage;
2. Network load balancing;
3. The virtual firewall.

## Server usage metric

With regards to the server usage metric, it is defined as "measuring using common pay per use metrics in IaaS and PaaS environments that are quantified by the number of virtual servers and ready-made environments" (Source: *Cloud Computing: Concepts, Technology & Architecture*, Erl, T., Arcitura Education, 2013, p. 389).

As it is defined, the server usage metric is divided into the following metrics:

1. The on demand virtual machine instance allocation metric;
2. The reserved virtual machine instance allocation metric.

The first metric measures the usage fees in the short term, whereas in the latter, the usage are measured can calculated over the long term:

1. The on-demand virtual machine instance allocation metric:
   a. It measures the availability of the virtual server instance;
   b. The measurement component is the virtual start date to virtual server stop date.
2. The reserved virtual machine instance allocation metric:
   a. It measures the cost for securing a virtual server;
   b. The measurement component is the reservation start date to the date of expiration.

## Cloud storage device usage

In terms of cloud storage device usage, it can be defined as

> the amount of space allocated within a predefined period, as measured by the on demand storage allocation metric. Similar to IaaS based-cost metrics, on-demand storage allocation fees are usually based upon short term increments. Another common cost metric for cloud storage is the I/O data transferred, which measures the amount of transferred input and output data.
>
> *Source: Cloud Computing: Concepts, Technology & Architecture, Erl, T., Arcitura Education (2013, p. 390)*

1. The on demand storage space allocation metric:
   a. It measures the time length as well as the size of the demand storage space;
   b. The measurement component is in terms of the date of the storage release and the reallocation to date of the storage allocation.
2. The I/O data transferred metric:
   a. It measures the amount of the data which is being transferred between the cloud computing infrastructure;
   b. The measurement component is in bytes.

Along with these usage metrics, there are also other cost management considerations which need to be described, and they are as follows:

1. Cloud service design/development: This is when the pricing models and the cost templates are defined by the ISP;
2. Cloud services deployment: This is when the usage measurement is ascertained and put into place in the cloud computing infrastructure;
3. Cloud service contracting: At this stage, the internet service provider determines the actual usage cost metrics;

4. Cloud service offerings: This is when the ISP formally publishes the actual costs for usage of the IT assets and the IT resources;
5. Cloud service provisioning: This is when the cloud service thresholds are established;
6. Cloud service operations: This is when the active usage of the cloud computing infrastructure results in actual cost usage metric data;
7. Cloud service provisioning: This is when a cloud computing infrastructure is decommissioned, and the cost data are archived and stored.

Now, that we have looked at some very important costing metrics, we not at this point in the chapter, turn our attention to the Quality Metrics which define a cloud computing infrastructure. These Quality Metrics consist of the following components:

1. Availability: This refers to the service duration of the cloud computing infrastructure;
2. Reliability: This refers to the minimum time between failures;
3. Performance: This refers to the response time, and capacity of the cloud computing infrastructure;
4. Scalability: This refers to the fluctuations in demand on part of the end user or the small business owner;
5. Resiliency: This refers to the how quickly the cloud computing infrastructure can be backed up and recovered in a time of disaster;
6. It must be quantifiable: This variable must be clearly established and absolute;
7. It must be repeatable: This variable is the ability to receive identical results;
8. It must be comparable: The units of measure (such as the amount of bytes) need to be standardized and made the same for all levels of comparison within the cloud computing infrastructure;
9. Easily attainable: This should be a common base form of measurement which can be most importantly used and understood by the end users and the small business owners.

The cloud computing infrastructure quality metrics are as follows:

1. Service availability metrics;
2. Service reliability metrics;
3. Service performance metrics;
4. Service scalability metrics;
5. Service resiliency metrics.

## Service availability

In terms of the service availability metric, it is composed of two submetrics:

1. The availability rate metric: This refers to the fact that an IT asset or IT resource can be expressed as a percentage of total uptime of the entire cloud computing infrastructure:
    a. It is measured as a total percentage of service uptime;
    b. The measurement component is in Total Uptime.
2. The outage duration metric: This particular metric measures both maximum and average continuous outage service level targets:
    a. It is measured in terms in terms of the duration of a single outage;
    b. The measurement component is the date/time of the start of the outage, as well as the date/time of the end of the outage.

The service reliability metrics can be defined as

> the reliability that an IT resource can perform its intended function under pre defined conditions without experiencing failure. Reliability focuses on how often the service performs as expected, which requires the service to remain in an operational and available state.
>
> *Source: Cloud Computing: Concepts, Technology & Architecture, Erl, T., Arcitura Education (2013, p. 407)*

As can be seen from the definition, the service reliability metrics consist of primary two metrics:

1. The meantime between failures:
    a. It measures the expected time between failures;
    b. It is measured in terms of internet service provider operating times and the total number of failures.
2. The reliability rate metric can be defined as "the reliability rate that represents the percentage of successful service outcomes. This metric measures the effects of non fatal errors and failures that occur during uptime periods"
   (Source: *Cloud Computing: Concepts, Technology & Architecture*, Erl, T., Arcitura Education, 2013, p. 407).
    a. This measures the actual, percentage of the successful outcomes under certain conditions;
    b. The measurement component is the total number of successful responses per the total number of requests.

## Service performance metrics

The service performance metrics "refers to the ability on an IT resource to carry out its functions within expected parameters. This quality is measured using service capacity metrics, each of which focuses on a related measurable characteristic of IT resource capacity" (Source: *Cloud Computing: Concepts, Technology & Architecture*, Erl, T., Arcitura Education, 2013, p. 407). This overall metric metric consists of the following specific metrics: "Cloud Computing: Concepts, Technology & Architecture", Erl, T., Arcitura Education, 2013, p. 407).

1. The network capacity metric:
   a. The measurement component is the amount of total network capacity;
   b. It is measured in terms of throughput (bits per second).
2. The storage device component metric:
   a. It is measured in terms of storage device capacity;
   b. It is measured in terms of gigabytes.
3. The server capacity metric:
   a. The measurement component is that of the server capacity;
   b. It is measured in terms of CPU frequency in GHz, and the RAM storage size in gigabytes.
4. The web application capacity metric:
   a. The measurement component is the web application capacity;
   b. It is measured in terms of the rate of requests per minute.
5. The instance starting time metric:
   a. The measurement component is the time which is required to start a new instance;
   b. It is measured in terms of the date/time of instance startup to the date/time of the start request.
6. The response time metric:
   a. This measurement component is the time required to perform synchronous operations;
   b. It is measured of the total number of responses divided by the total number of requests.
7. The completion time metric:
   a. The measurement component is the time which is required to start and finish an asynchronous task;
   b. It is measured by the date of requests and responses divided by the total number of requests.

## Service scalability metrics

With regards to the service scalability metrics, they can be defined as "metrics which are related to IT resource elasticity capacity, which is related to the maximum capacity that an IT resource can achieve, as well as measurements of its ability to adapt to workload fluctuations" (Source: *Cloud Computing: Concepts, Technology & Architecture*, Erl, T., Arcitura Education, 2013, p. 409). As it can be seen from the definition, it contains the following metrics:

1. The storage scalability:
    a. The measurement component are the required changes to higher and more demanding workloads;
    b. It is measured in gigabytes.
2. The server scalability (horizontal):
    a. The measurement component is the server's capacity changes in the face of increased workloads;
    b. It is measured in terms of the number of virtual servers which are available.
3. The server scalability (vertical):
    a. The measurement component is the virtual server spikes;
    b. It is measured in terms of gigabytes.

## Service resiliency metrics

In terms of the service resiliency metrics, it can be described as

> the ability of an IT resource to recover from operational disturbances is often measured using service resiliency metrics. When resiliency is described within or in relation to SLA resiliency guarantees, it is often based on redundant implementations and resource replication over different physical locations, as well as various disaster recovery systems.
>
> *Source: Cloud Computing: Concepts, Technology & Architecture, Erl, T., Arcitura Education (2013, p. 411)*

This type of metric can occur at all phases of the cloud computing infrastructure:

1. The design phase: This is how well prepared the cloud computing infrastructure is to cope with challenges and disasters;
2. The operational phase: This refers how well the differences in the levels of service agreements during and after a serious downtime or even a service outage'
3. Recovery phase: This refers to the rate at which the IT assets and the IT resources can recover from downtime.

As it can be seen from the definition, there are two types of primary metrics here:

1. The meantime to switch over metric:
   a. The measurement component is the expected time to complete a virtual server switchover from a failure to another virtual server which is located in a totally different geographic region;
   b. It is measured in terms of switch over completion divided by the total number of failures.
2. The mean time system recovery metric:
   a. The measurement component is the expected time to perform a complete recovery from a disaster;
   b. It is measured in terms of time of recovery divided by the total number of failures.

## AN INTRODUCTION TO BIOMETRICS IN THE CLOUD

Now that we have an extensive overview of what the cloud computing infrastructure actually entails, in this section, we will examine what biometrics in the cloud will look like. As previously mentioned at the beginning of this article, there are already a number of biometric applications which are using the cloud infrastructure in some shape or another.

But, these are not full blown, or complete implementations, but rather, they are just a micro-application (in other words, just using some aspect of the cloud). Also, the biometric technologies utilized are very limited in nature-primarily just voice recognition is being used.

The definition of biometrics in the cloud means that a full blown biometrics infrastructure will be placed at the hands of the hosting provider. Also, the type of application (and even the number of different kinds of applications) will be unlimited in scope, and this hosted biometrics infrastructure will be designed to fit the wide gamut of biometric applications which exist today and those which are under research and development.

This includes the biggest market drivers of physical access entry; time and attendance; single sign on; large-scale law enforcement (in particular AFIS); the national ID card; the e-passport; as well as other types and kinds of verification and identification scenarios.

Also, this hosted biometrics infrastructure will be designed to support any type or kind of biometric technology which is available today, as well as those which are also under research and development. This includes the physical biometrics (fingerprint recognition, face recognition, iris/retinal recognition, hand geometry recognition, and voice recognition) and the behavioral biometrics (keystroke recognition and signature recognition).

The concept of biometrics in the cloud will also be catered extensively to one very important component-the customer, especially that of the small business owner. Anybody or any organization or any business, no matter how large or small they are, or wherever there are in the world, will be able to use cloud-based biometrics at any time that they are considering of implementing a biometrics infrastructure at their place of business.

Therefore, biometrics in the cloud can be envisioned from two very different aspects: (1) the hosting provider; and (2) the customer, or the small business owner. Both parties will have their own responsibilities in order to ensure that biometrics in the cloud does indeed become a reality.

First, let us look at this from the standpoint of the small business owner. Imagine if you will that a small business owner wishes to implement a biometrics-based security system and is giving heavy consideration to the cloud-based option. Just exactly how will the small business owner go about this process? Well, we have to first assume that the small business owner has no prior knowledge of biometrics what so ever. Thus, it is safe to assume that they will have to their own level of research of what needs to be done, and most likely, will have hired a biometrics consultant to explore all of the alternatives and possibilities of what can be done. This part will have included a major systems analysis and design of the small business owner's existing security infrastructure, and exactly what and where the role biometrics can take place.

Next, the choices of what to do next will come down to two items: (1) whether or not to implement biometrics as part of the existing security infrastructure – or – to deploy it as a brand new implementation; and (2) what type of biometric technology to utilize. The answer to the first question will of course depend primarily upon the results of the systems analysis and design just conducted. But with the second question, with a cloud-based deployment, any biometrics technology should work.

But there are also other important factors which need to be taken into account with this, such as ease of use, ease of deployment, ease of end user training, as well as cost affordability given that the small business owner will most likely be on a stringent budget. With these variables taken into account, vein pattern recognition will be the choice of technology to use for biometrics in the cloud and is further discussed in the next section.

Now, that the small business owner has chosen to utilize vein pattern recognition as their choice of technology for biometrics in the cloud, the next question is where to have the actual hardware devices installed at their respective place of business. Again, this is 100% dependent upon the application which will be utilized.

For example, if it is for physical access entry, the biometric devices will be obviously contained at the major access points at the business or organization, if it is for time and attendance, then the devices will need to be installed near where the employees work, or if it is for single sign on (which is for eliminating the use of passwords), then the biometric devices will have to be installed at each of the employee's workstations, and so on.

It is most likely that the small business owner will have the actual biometrics Vendor come out and install whatever the number of vein pattern recognition devices is required. Now, this is where the beauty of biometrics in the cloud comes into play. For example, under traditional deployments, installing and implementing a full-fledged biometric system can take a lot of time, effort, and exorbitant costs, especially on part of the customer, if it is a large organization.

And, this is especially true if the biometric system is to be integrated into an existing, or a legacy security system infrastructure at the place of business. But, with biometrics in the cloud, of the biggest advantages is that a biometric system can be implemented either into an existing, or legacy security infrastructure as well as a brand new deployment with the greatest of ease.

For instance, all the small business owner has to do is merely acquire the vein pattern recognition devices, install them, and also ensure that the proper networking is configured correctly so that these devices can access the cloud-based biometrics infrastructure from wherever it is located in the world.

After all of the vein pattern recognition devices have been installed, and the proper networking has been ensured, the next major step for the small business owner is to now activate their respective account with their biometrics in the cloud infrastructure plan with the hosting provider. It is envisioned that this will happen in a manner which is very similar when one chooses and activates a new plan with an internet services provider, for such services as web hosting, e-mail, internet telephony, and backup storage.

Of course, once the account is set up and activated, then the small business owner can enter all of the names of the employees who will be using the system, the type(s) of application(s) for which biometrics in the cloud will be used for, and all of the other relevant services which come with a traditional biometrics deployment. This includes such as things as verification/identification services, template reporting and transaction services, administration services, and the relevant software applications, and much, much more.

It is anticipated that from when the time the devices are actually installed and activated, it will take just a matter of minutes until a full blown, cloud-based biometrics application will be made available to the small business owner. Again, this is one of the major benefits of biometrics in the cloud. For example, because of the hosted nature, a complete biometrics implementation and application can be set up in just a matter of minutes, as opposed to the days or even weeks with the traditional methodologies of biometric system deployment and implementation.

But an important point must be noted here. It is assumed that the small business owner will have established a relationship with a hosting provider and have the account setup before the actual devices are installed. In the beginning, there may not be a lot of hosting providers offering cloud-based biometric services, thus the small business owner may be very limited in their choice of where their biometrics infrastructure will be hosted at. But it is anticipated that over time, and as the demand for biometrics in the cloud increases, the number of hosting providers will proliferate much like that on the level of ISPs.

Now, that we have examined the concept of biometrics in the cloud from the stand point of the end user, most notably, the small business owner, it is equally important, if not more, to look at all of this from the perspective of the hosting provider. From this standpoint, one of the first questions to be asked will be is if the existing ISPs will take on this new role of hosting a biometrics infrastructure, or will brand new ISP's have to be created and dedicated to just doing this?

Without conducting an extensive benefit-cost analysis, it will be very difficult to answer this question in detail. But at first glance, it would make much more sense for the existing ISP's to take on this added role, at least initially. After all, the necessary hardware is already in existence, and at the beginning stages, it is anticipated that only software modifications and enhancements will have to be done to support a cloud-based biometrics infrastructure.

Another point which needs to be taken into consideration is that existing ISP's are not simply going to jump on board to support biometrics applications. After all, this is a huge change in the way they currently do things, and this will incur changes in the existing processes which will no doubt require deep levels of testing before anything goes live to the end user.

In this regard, it will be very crucial for the biometrics vendors as well as the industry as a whole to "make their case" to these ISP's, and to work with them very closely in order to ensure that the overall biometrics in the cloud infrastructure will be launched successfully and smoothly.

In other words, strong partnerships and alliances have to be formed between the ISP's and the biometrics vendors. Once a handful of leading ISP's have been identified, the next step will be for these ISP's to take on now the added role of hosting a cloud-based biometrics infrastructure. Thus, it will take crystal clear communications and very clear planning in order to ensure this transitory step occurs as smoothly as possible.

Also, here in lies another strategic advantage of biometrics in the cloud. For example, as this infrastructure starts to roll out with the ISP's, except for the software modifications and some minor hardware changes, this concept will already assume the existing cloud infrastructure described earlier in this section. This infrastructure includes PaaS, SaaS, and IaaS.

In further analyzing these three components which will support a biometrics in the cloud infrastructure, it is crucial at this point to take a top down approach. Thus, in this regard, it is the IaaS which can be viewed as the overarching layer in this model. In fact, one can consider the IaaS to the underlying architecture for which the next two components of SaaS and PaaS will be built upon. The IaaS consists of the actual servers, storage (including the databases), the networks, and the other major computing resources. So as you can see, it is the IaaS which as the very heart of biometrics in the cloud.

But before we delve further into how the subcomponents of the IaaS will support a cloud-based biometrics infrastructure, one very important needs to be made here. Although the actual servers and other relevant hardware are contained at the ISP, the actual raw storage units are further partitioned into smaller subsections from within the hard drive of the server.

These smaller subsections are referred to what is known as "virtual servers". In other words, these virtual servers contain all of the software resources the actual, physical server has, thus giving the small business owner the look and feel of having their very own server from within the control panel of their hosting account. As a result of this virtual server, biometrics in the cloud thus possesses one of its strongest advantages. And that is, the low, predictable, and monthly fixed costs which makes this whole cloud infrastructure very affordable to the small business owner.

The IaaS component would consist of the following features, which in turn, would provide the underlying architecture for biometrics in the cloud:

*Virtual servers:* This is the main point of "contact" of you will, where the small business owner will have total control over their biometrics in the cloud infrastructure. At this level, the small business owner will have a control panel (much like a web/e-mail hosting control panel) from which he or she will be able to oversee all transactions and have total administrative control over their cloud-based application. The underlying premise here is that the customer can essentially do whatever they need to or desire to for their cloud-based biometrics infrastructure without intervention. But, the customer will not be able to do any kind or type of software upgrades or modifications to their virtual server, as it will be up to the hosting provider to do all of this. The key component of the virtual server is the operating system, and this will support both open source and closed source platforms (such as Windows and Linux, respectively), as described below.

On another note, should the small business owner require technical assistance of any kind, support will come from two sources: (1) the hosting provider itself (for technical assistance with the underlying hardware and software); and (2) the biometrics vendor(s) who are working with the ISP (in this case, the biometrics vendor(s) will provide technical assistance in terms of the biometrics perspective in these cloud applications).

*Storage:* This will be the place where all of the biometric templates will be stored, as well as the associated databases which will process the various transactions of these templates. In this regard, there are several key points to be made here. First, the small business owner will have full development freedom in the way and manner in which they wish to design and create their required databases. Meaning, they will have a choice over the technology platform from which upon the respective databases will be built.

The small business owner will have their choice if they want to utilize either a closed source or open source database model. The former refers to UNIX-based architectures such as MySQL and PostgreSQL, and the latter refers to Windows-based architectures as SQL Server and other Oracle-based architectures. Of course, closed source databases will mean an extra expense for the ISP in terms of software licenses and upgrades. Thus, the small business owner will have to pay the respective fees to the ISP in order to contain their costs.

In these databases, the biometric templates to be stored will include the enrollment templates and the verification templates of each employee the small business owner has. Of course, depending upon the development which is done, robust statistical and reporting features will be made available to the small business owner as well, so that he or she can analyze the effects and impacts of their cloud-based biometrics infrastructure.

Also, here in lies of the other great advantages of biometrics in the cloud: scalability. For example, in traditional biometric deployments, the databases have to be literally redesigned and rebuilt in order to cope with the demand for increased scalability.

But, with the cloud, the biometrics databases(s) can be scaled to fit any array of biometrics applications within just a matter of minutes for a wide array of applications, ranging from the simplest 1:1 verification scenarios to the most complex 1:many law enforcement identification applications.

*Networks:* This is the third major component of the IaaS. Under this, a biometrics in the cloud infrastructure will have network connectivity to the vein pattern recognition devices at the physical location of the small business owner and vice versa. But, in order for each side to recognize each other (the hosting provider and the vein pattern recognition devices), each cloud-based biometrics application will have its own unique internet protocol (IP) address so that the small business owner and their respective cloud-based infrastructure can be uniquely identified in the vast expanse of the Internet.

Also, given the shared choice of open and closed source software platforms as described above, the appropriate networking protocols will also have to be offered by the ISP to the small business owner. But, it should also be kept in mind that the networking component will be one of the most critical features of biometrics in the cloud. This is so, because the two second or less verification times which have been realized in traditional biometrics applications will also have to be realized in cloud-based applications. This is especially important in the case if massive, law enforcement identification applications are applied to the cloud as well. Gargantuan searches through millions of biometric template records will have to come down to just a matter of minutes.

Thus, as one can see, network and bandwidth optimization will of upmost importance, as well as the processing power of the virtual servers. Also, there is the question of just how safe the biometric templates are when they are in transit across the network.

For example, do they need to be encrypted? To allay these concerns and fears for the small business owner, virtual private networks will also be offered. With this, the data packet which stores the biometric template will be further encapsulated by another data packet, thus providing greater assurances of security.

The second major component of the biometrics in the cloud infrastructure is the SaaS. In the cloud world, generally speaking, SaaS is fast becoming a popular choice. And in biometrics, it too is becoming very popular. But with biometrics in the cloud, the SaaS component will come from two different angles: (1) software which is already created and can be purchased on demand by the small business owner; and (2) a software platform in which full, customized development will take place.

Keep in mind, at this point, it is the SaaS component in which the small business owner will come into contact the most in terms of their hosted biometrics infrastructure. Therefore, just like the networking aspect as described previously, the SaaS also has to work very smoothly with almost 100% uptime (all of the relevant uptimes for biometrics in the cloud will be spelled out in the SLA agreement).

With respect to the first software option, it is envisioned that the biometrics vendors will create and develop software applications for all of the biometric technologies which are available today, both in terms of physical biometrics and behavioral biometrics. The applications developed will fir all of the major market categories including physical access entry, time and attendance, and single sign on.

Of course, these applications will not be developed for free. The biometrics vendors will sell these to the ISP's, and provide the relevant support which would be needed. In turn, the ISP's would make these software applications, available on demand, to the small business owner for the fixed, monthly fee. It could also be the case that these software applications could also involve some certain degree of customization, in order to better fit the needs of the small business owner.

With regards to the customized software platform, this will give the small business owner the opportunity to create and develop their own biometrics applications. It should be noted that the ISP will not provide the development, rather, the small business owner will have to hire their own developers. In this aspect, both open source and closed source platforms will be provided.

With the open source model, probably the most popular development tool will be that of PHP, Perl, etc., and with the closed source model, the most widely used tool will be that of ASP.Net, as well as other software and content packages such as SharePoint. At this stage, it is envisioned that the biometrics software applications created will be web-based.

After all, it is the Internet which will be what the small business owner used to access their cloud-based biometrics infrastructure. Thus, web server software such as Apache (for open source) and Internet Information Services (IIS, for closed source will also be offered.

Also, with any software applications comes the backend – namely the databases. So, the IaaS and SaaS levels will have to work closely with another. With regards to the SaaS component, there will be one thing which will be very important. And that is having a set of common standards and best practices for software development of applications for biometrics in the cloud. Currently in the biometrics world, there is no central authority which oversees the many individual and disjointed biometrics standards and best practices which do currently exist globally.

Also, at the present moment, strictly speaking, the cloud in general is a huge nebulous of virtual servers, TCP/IP addresses, networks, and literally hundreds and thousands of applications, all coming together in very much resembling a jagged puzzle. Therefore, as biometrics in the cloud does over time proliferate in terms of adoption on a global scale, the need to have a central authority to oversee and enforce a listing of best practices and standards will become ever so important, with some substantial advantages to be gained.

For example, the small business owner can look up to this listing as a benchmark and guide as to how their biometrics in the cloud software applications should be developed. But, the biggest benefit will be that this open dialogue (as it is envisioned) will foster a sense of collaboration and support throughout the biometrics community, worldwide.

The third major component of the biometrics in the cloud is PaaS. The PaaS itself is not a unique structure liked the SaaS or the IaaS, because it is still a vehicle for software development. But the main difference is the magnitude upon which these biometrics software applications can be built.

For example, the SaaS will support on demand software applications and customized software development, as described previously. With the latter, the intention is to create those types of biometrics software applications which can support the needs of the small- to medium-sized businesses and organizations. The best examples of this are the one to one verification software applications.

But, where would the much larger-scale one to one verification and the one to many biometrics software applications reside? It would be in the PaaS. In other words, the PaaS can be viewed as an extension of the SaaS component from which much more complex software applications can be created and supported.

Just like the SaaS, the PaaS will also support both closed source and open source software development models. But the main difference between the two is that the PaaS will be much more exhaustive in terms of development resources to support the much more complex biometrics software applications.

There are a number of large-scale, biometrics applications that would not just be a candidate as a cloud-based application but would reside very well within the PaaS:

The national ID card: There are various nations around the world which have started to use this, and it is a method for citizens of a particular country to prove their citizenship. However, the national ID card has become much more than its original, intended purpose. It has become a document of sorts for which it can be used for e-voting, filing medical insurance claims, claiming government-based entitlements and benefits, etc.

But it is not a just a card – it is a smart card which can contain many biometric templates, ranging from fingerprint recognition templates all the way to facial recognition templates. As a result, this biometrics aspect of the national ID card could fit very well into the PaaS structure, because large-scale software applications would have to be developed to support the millions upon millions of the people of a nation who use it.

The passport: This is now taking the role of the traditional paper passport, for the identification of a foreign traveler. Just like the national ID card, the e-passport is really nothing but a smart card with biometric templates stored in it (and the storage space is quite high). In order to use the e-passport, the passenger merely has to point their e-passport to a reader (via a wireless communications channel), and if 100% verified, the traveler is allowed to enter the country of their final destination. Of course, the traveler has to submit their biometric templates at their country of origin. Because of this biometric aspect to the e-passport, it would be a perfect application for a biometrics in the cloud application, as well as for the PaaS, just given the sheer number of biometric transactions which have to occur each second in every airport around the world.

Law enforcement: In the United States, it is the AFIS database which has supported law enforcement needs at both the state and federal levels. AFIS stands for "Automated Fingerprint Identification System" and is a gargantuan fingerprint template database of known suspects, criminals, terrorists, etc. There are current efforts underway to update this legacy database. Also, at the same time, the technology associated with law enforcement has advanced greatly, and it is at the point now which wireless devices are used by law enforcement officials to collect the fingerprints of an apprehended individual in real time. This is then transmitted to the AFIS database to confirm the identity of the individual in question. But, there are time delays and processing wait times for these biometric transactions to take place, and the subsequent results presented to the law enforcement officer. Because of this, the AFIS database, if migrated to a biometrics in the cloud infrastructure, would drastically reduce these transaction times from minutes to seconds. Also, the massive software applications which are needed at the present time to support the AFIS database will fit very well into the PaaS component of cloud-based biometrics applications.

It should be noted at this point that these large-scale, biometrics applications, as well as others, cannot suffer any downtime what so ever, both from a network connectivity and infrastructure standpoint. As a result, redundancy is very important here, and in the case of any outage, all biometrics in the cloud resources would simply be rolled over to the backup IaaS, PaaS, and SaaS. This is one of the other great advantages of biometrics in the cloud. Given its pooled and shared resources nature, redundancy is much more possible, as opposed if these infrastructures were physically housed within the entities themselves.

Now that the major concepts of the biometrics in the cloud infrastructure have been reviewed, the next major question to be looked at is what kind of cloud model should be offered. Four different types of cloud models were discussed earlier in this section. Thus, given that the premise of the biometrics in the cloud infrastructure is to be made available to any entity, any organization, at anytime, anywhere in the world, and on demand, the public cloud model will work best.

The only criterion that will be used in this cloud model is if the business entity or organization has the financial means to purchase the desired or needed cloud-based biometrics applications. Today, the primary cloud infrastructure offered by the various ISP's is that of the public model. The only discrepancy is that of pricing. Some IPS's offer free services, some offer services at exorbitant prices, and some are middle of the road of the pricing scheme.

Therefore, when choosing the ISP for their cloud-based biometrics infrastructure, the small business owner should remember the old adage "You get what you pay for". But also, it should be kept in mind that that the ISP's should not be charging exorbitant pricing either, and this will have to be carefully monitored. One of the ultimate goals of biometrics in the cloud is to have the ability to offer small business owners affordable levels of service (as well as enterprise grade), as opposed to the high cost of traditional biometric deployments.

## A detailed example of how biometrics in the cloud would look like

This section will now put both together into a real world example of what biometrics In the cloud will look like. Once again, the two major components throughout this entire process is: (1) the small business owner (or really any other customer for that matter); and (2) the hosting provider in which the biometrics in the cloud infrastructure would reside at. First, let us begin this illustration from the standpoint of the small business owner.

It is safely assumed that the small business owner has some sort of security need – thus, they are exploring a cloud-based biometrics solution for their needs. After conducting a comprehensive requirements analysis (of course, with the help of a biometrics consultant), the small business owner decides to implement a cloud-based biometrics solution.

Furthermore, it has been decided will be a brand new security solution. Meaning, it will not be part of an existing security infrastructure, rather, this new cloud-based solution will be a primary means of security, not depending upon other legacy security systems. This new solution will be used for physical access entry, in which the small business owner will allow employee entry into the place of business strictly based upon their ability to be successfully verified by the biometrics system.

Also, it is assumed in this scenario that the small business owner has cost considerations; would like to have a biometrics system which will be easily an readily accepted by his or her employees; the employee training in learning how to use the system will have a very low learning curve; and have a system which can be very easily and quickly installed, imposing no further impediments to existing business processes.

With all of this in mind, the small business owner ultimately chooses vein pattern recognition, and they have also picked out the supplier who will come and install the hardware, as well as the requisite networking which will be needed for the vein pattern recognition hardware to link up with the cloud-based biometrics infrastructure located the hosting provider.

For simplicity of this illustration, assume that the small business owner decides to install just one vein pattern recognition device at the main point of entry at the place of their business, and that the number of employees is small – say about 20 employees. All that is needed at this point is to have the actual vein pattern recognition device installed, the networking completed, and the employees have been trained in how to use the system.

But, there is one more step which needs to be completed after all of this, and that is the setup of the biometrics in the cloud-hosting account, and the infrastructure configuration. This part will be examined in more detail after the second piece is now examined – the hosting provider.

From the standpoint of the hosting provider, the biometrics in the cloud infrastructure will actually be quite simple when compared to the many other complex applications which can be hosted. Since the business is small, the application needs will be small as well. For example, in this scenario, all that will be needed is 1:1 verification.

So, as a result, the only cloud components which will be needed are the IaaS and the SaaS. But it should be noted at this point that that as the security needs of the small business owner grows and expands over time, more components and services from the cloud-based biometrics infrastructure can be added on demand.

For instance, should the small business owner drastically expand their business into a much larger one, to include applications which involve 1:many identification scenarios, the subcomponents of the PaaS can be implemented within just a matter of minutes. Likewise, the opposite is also true. Should this small business owner decide to cut back on the components and services for which they originally signed up for, these can be removed on demand as well.

Looking at the standpoint from the IaaS, the small business owner will need the following:

- *The virtual server:* This will be the heart of the biometrics in the cloud infrastructure. It is at this point that the small business owner will have total control over their entire infrastructure.
- *Storage:* This part will consist of the database upon which the biometrics templates will be stored at. Since the needs are quite simple, only one database will be required. The small business owner, of course, will have their choice of technology from which to build their database, but it is safe to assume at this point that the initial design will be kept quite simple. Once the virtual server is configured, and the employees have started to enroll into the vein pattern recognition reader, the first types of data to be stored in this database will be the enrollment templates.
- *Networking:* The importance of this subcomponent cannot be emphasized enough. It is essential that the networking is working at peak performance every day and every second. For this example, the standard networking protocol of TCP/IP will be used (this will be the default protocol used, unless the small business owner specifies otherwise) to communicate between the vein pattern recognition device and the cloud-based biometrics infrastructure.

Looking in terms of the SaaS component, this is what will be needed:

- *Software:* This will be another, subsequently important piece for the small business owner once everything is all set up. It should be noted at this point that the IaaS will have to be configured first before the SaaS subcomponent can be established. Given such simple needs in this example, the question often arises if any software application is needed at all. But, let us assume at this point that the small business owner would like to have a software application which is designed for vein pattern recognition, and for 1:1 verification scenarios. By having this, the small business owner can see first hand how well their biometric system is actually working. For example, they can robust statistical reporting features, which can help them optimize the system even more. Also, because of these simple needs, the small business owner decides to go with a software application which is already pre developed, therefore, there is no need for any customized or specialized software development.

Now, that both of the sides have completed their side of the setup (remember, the hosting provider will always have their biometrics in the cloud services ready literally, 24 × 7 × 365), the two now have to be linked to each other.

This linkage will be completed through the control panel that the hosting provider has given to the small business owner. Through this control panel, during initial configuration of the virtual server, the small business owner will be guided through a series of steps which will lead them to the ultimate setup of their biometrics in the cloud infrastructure.

The following are some of the major steps the small business owner will have to complete:

1. *The type of application to be created:* The small business owner will select either physical access entry; time and attendance; single on, etc.
2. *The type of biometrics devices installed:* The small business owner will select from a list of choices of what types of biometrics they have installed to support their cloud-based biometrics infrastructure; as well as the total number which is installed;
3. *The type of matching services needed:* The small business owner will select if they need 1:1 verification or 1:many identification matching services for their application;
4. *The number of employees or people who will be enrolled in the system:* The small business owner will have to enter how many people will be enrolled thus far;
5. *If any software applications will be needed:* The small business owner will be asked to select if they want to have a pre-developed software application or develop one on their own. Should they choose the former, it will be ready to go immediately after the configuration has been completed;
6. *If any databases will be needed:* The small business owner will either select yes or no, if they select yes, they will then be presented with a list of database technologies they choose from in order to create their database(s).
7. *Agreement to terms of service and contract; credit card information entered*: Before the cloud-based biometrics infrastructure, the small business owner has to agree to abide by the contract, and credit card

will be the primary means of payment (of course, they will have various payment options as well, such as monthly and yearly).

Once all of the above has been completed, and the small business owner finally clicks on the "Submit" button, the biometrics in the cloud infrastructure will now be set up and ready to go. However, a number of key point have to me made here. First, the above steps are just a sampling; it is quite conceivable that many other options could be presented to the small business owner during this process.

Second, the hosting provider will require the small business owner to submit a series of test biometric templates after initial setup, in order to ensure that all is running smoothly and without any errors before going live. Third, this biometrics in the cloud control panel will also have another area from which the small business owner can select the other services and add-ons, as needed (such as database setup/configuration; technical support; running certain types of tests on prototype applications; and software development/configuration).

Fourth, many ISP's offer their customers certain web hosting packages to their customers, which fits their needs and budgetary constraints. But, with the biometrics in the cloud infrastructure, this will be different. Meaning, once the small business owner has activated their account and completed the initial virtual server configuration, they will have access to all of the services and options, right from the beginning. In other words, the biometrics in the cloud infrastructure will be all inclusive, and there will be no different packages to choose from.

The small business owner can always upgrade or downgrade their services and options on demand and will only be charged for the services they have rendered. Fifth, with this all inclusive nature, the small business owner can host many other different biometrics applications from within this one infrastructure.

## A review of the advantages and the disadvantages of biometrics in the cloud

At this point, now that the major concepts of biometrics in the cloud have been examined, and an actual example provided, it is important now to review the major advantages and disadvantages of biometrics in the cloud.

The advantages include:

1. *An instant setup of a biometrics infrastructure:* Within literally minutes, this entire system can be set up, all at the click of a mouse. The only exception to this is the time delay which will be experienced when the biometric devices are installed at the place of business or organization;
2. *It is on demand:* The biometrics services and other components can be added on or cancelled instantaneously;

3. *It is affordable:* This is especially the case for the small- to medium-sized businesses. Because of the cloud-based infrastructure, any costs will be available at a fixed and predictable monthly price; this is unlike traditional biometric systems where the costs can greatly escalate over a short period of time;
4. *It is highly scalable:* With this in mind, a biometrics application can be cut back or expanded greatly in just a matter of seconds; with the traditional biometric deployments, and upgrades or downgrades can mean an entire revamp of the whole system, thus making it virtually cost prohibitive to do;
5. *An entire community of technical support and collaboration:* Given the open source nature of biometrics in the cloud, and that a set of best practices and standards will have to be adopted, customers from all walks of businesses and enterprise can readily share their experiences and expertise to others who are contemplating of launching a biometrics in the cloud infrastructure;
6. *The business owner has no responsibility for hardware or software:* Except for the cost of the biometric device hardware, the small business owner assumes no more responsibility for this; this is the ultimate responsibility of the hosting provider for upgrading the hardware and software for the biometrics in the cloud infrastructure, including software licensing fees as well as maintenance costs;
7. *Interface with legacy systems:* A cloud-based biometrics infrastructure can fit quite easily and readily into an existing security infrastructure should the small business owner choose to have this option (in other words, a cloud-based biometrics application would provide for a superb multimodal security solution);
8. *An existing infrastructure in place:* For those ISP's which will be offering a biometrics in the cloud infrastructure, the transition should prove to be a relatively smooth one; this is so because all of the existing hardware and software that the ISP has will be used to create a biometrics in the cloud infrastructure;
9. *Redundancy is very easy:* Given the pooled and shared resource nature of the biometrics in the cloud infrastructure, redundancy is thus very easy and very cost effective; in traditional biometric deployments, redundancy very often means extra servers to store the biometric templates and other processes, which mean a much greater expense;
10. *An enterprise grade level biometrics system is available to all:* Traditionally, those biometric systems with the best performance and of peak sophistication were only available to those businesses who could afford it-namely the Fortune 500 companies. Now, with biometrics in the cloud, an enterprise grade biometrics system can be made available to all, and not just a select few.

11. *New cloud-based services could emerge:* For example, forensics makes heavy use of biometric evidence which is left behind at a crime scene and is collected later by experts. The most common type of this evidence is what is known as "latent fingerprints". Therefore, an offshoot of biometrics in the cloud could very well be what is known as "Forensics as a Service".

The disadvantages include:

1. *Time:* This is probably the biggest disadvantage for biometrics in the cloud. For instance, biometrics has a very slow adoption and acceptance rate, especially here in the United States and in Europe;
2. *A long lag time for the ISP's:* The ISP's also will have a very long time to "warm up" to the idea of offering a cloud-based biometrics infrastructure. For example, many ISP's today just offer hosted office-based solutions for their customers, which is where the demand is right now at the present time. It will take a lot of work and convincing on part of the biometrics Industry to prove to the ISP's why they should offer a biometrics in the cloud infrastructure for their customers, as well as the positive return on investment (ROI) they should experience over the course of time;
3. *Changes in ISP business processes:* Although it is expected that the cost of any extra hardware or software imposed upon the ISP's should be minimal, it is also expected that the ISP's will be very hesitant from the outset about implementing biometrics in the cloud; after all, this means a drastic shift from the current business processes, because it is just human nature in general to be very hesitant about positive change in general.
4. *Changes in biometric business processes:* At the present time, each and every biometrics vendor has their own production processes, research and development strategies, and own corporate objectives. For instance, some vendors offer differing kinds of products and services. But, when biometrics in the cloud does indeed become a reality, the entire biometrics industry as we know it will have to shift their strategies to an almost services-based environment.
5. *A cloud-based biometrics infrastructure needs to be scalable to demand:* When biometrics in the cloud does indeed become a reality, it is very important that the ISP's not devote entire resources just yet to cloud-based biometrics. Meaning, there should not be idle cloud-based resources expecting to be consumed, when they may never be. Rather, resources should be expended which is proportionate to the demand for biometrics in the cloud. Meaning, there is a positive correlation between resources available and demand cloud-based biometrics. It has been determined that 85% of idle resources in the

cloud is totally unacceptable. Also, the biometric algorithms need elasticity in order to fit the scalability requirements of the cloud.

6. *Legalities and Privacy Rights could become grave issues:* Biometrics is one of those technologies which no matter what, it is always prone to claims of violations of civil liberties, loss of personal freedom, etc. This fear will only be heightened more as biometrics moves into the cloud, because the biometric templates will be literally held at the hands of the hosting provider. Also, grave legal concerns could take place to establish the clear separation between the biometric templates and other information/data which is stored at the hosting provider, even though technically they may be separate as they will be stored on different servers.
7. *Small-scale biometrics applications are best suited:* It has been determined at the present time that small-scale biometric implementations are much better suited as a cloud-based offering than the much larger-scale biometric applications. The former includes such things as the template creation process and verification scenarios. The major culprit inhibiting the growth of large-scale biometrics in the cloud is that the mathematical algorithms are not yet optimized for on-demand, elastic virtualization.

Chapter 6

# Biometric applications of the future

## BIOMETRICS GOING INTO THE FUTURE

As the world continues to further evolve, it is highly anticipated that biometrics will not only advance in terms of the technology, but it could even find new markets applications, other than the traditional ones that were reviewed earlier in Chapter 1. Furthermore, it is also expected that as the digital transformation also further evolves as well, the acceptance of biometrics among the world's population will also grow as well.

### E-voting

This will probably be most realized in the developing nations, primarily those in Africa and parts of Asia. One the catalysts for this is that these countries are now adopting digital-based technologies at a fast pace. For example, the ability to own and use a smartphone, once thought impossible just a decade ago, has now become a reality for most of the citizens in these countries. This is driven by the increased investments made by the respective governments into a robust and reliable network infrastructure.

Another key reason for the rapid adoption of biometrics in these developing nations is that now a majority of them have the right to vote. As a result of this, these governments are also investing heavily into a digital voting infrastructure. For example, the days of casting a paper ballot will soon be dissipated, as many of these citizens can now vote from their place of residence. Or, if they are on the go somewhere, they easily vote from the easy access of their smartphone.

But probably one of biggest reasons why biometrics will see such a huge uptick in these developing nations is that it now gives the citizens one of the greatest opportunities of actually being recognized in the eyes of their own government. For the longest time, many of the leaders in these developing nations for the longest time have ignored their own very own citizen based upon the fact they never even knew that they existed.


DOI: 10.4324/9781003562146-6

Because of this, a majority of these citizens could not even receive the most basic of entitlements and benefits that were allowed to receive. But now, with the advent biometrics coming into their landscape, these very people can now for once be counted and fully recognized. The primary reason for this is that the Biometric Templates that have been created for them are now irrefutable pieces of evidence that they actually exist. And because of their high level of uniqueness that is involved, the governments of these developing nations cannot claim as easily that fraud has been perpetrated when these people claim their identity, and say who they are.

## Multifactor authentication

But in the end, it is expected that the role of biometrics may be used for a multimodal approach, rather than using them directly as the sole means of authentication and authorization. One of the best examples of this is of multifactor authentication, also known as "MFA". This is where more than one authenticating mechanism is used to authenticate the end user in question. In this kind of methodology, at least three or more differing authenticating mechanisms must be used. For example, this can include a Challenge/Response Question, an RSA Token, and a Biometric Modality, such as that of Fingerprint Recognition and/or Iris Recognition. This can be seen in the illustration below:

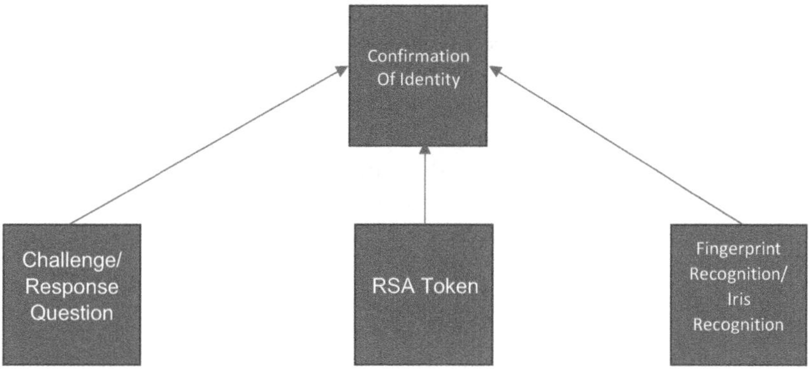


It is important to note that the above is just an example of the authentication mechanisms that can be used. As just reviewed previously, it is quite likely that even up to four or five differing authentication mechanisms could be used as well, but the common denominator still remains that the use of Biometric Technology will play a very dominant role in this scenario. It should also be noted MFA is always used when an access to something is typically needed, as that getting of getting resources from a server.

## The zero trust framework

Another trend that is occurring that will make heavy usage of biometrics is what is known as the "Zero Trust Framework", also known as the "ZTF" for short. A technical definition of it is as follows:

> Zero Trust is a security framework requiring all users, whether in or outside the organization's network, to be authenticated, authorized, and continuously validated for security configuration and posture before being granted or keeping access to applications and data. Zero Trust assumes that there is no traditional network edge; networks can be local, in the cloud, or a combination or hybrid with resources anywhere as well as workers in any location.
>
> *Source: What is Zero Trust Security? Principles of the Zero Trust Model | CrowdStrike*

To put in simpler terms, the ZTF is where the IT and network infrastructure of a business is actually broken out into different segments or "zones". Each of these segments (zones) is separate from one another, and in order to access a particular one, the end user must pass through an MFA checkpoint, and must present at least three (or even more) differing factors of authentication before they can gain access to any of the shared resources that reside in it.

One of the basic objectives of the ZTF is to make it statistically almost impossible for the cyberattacker to break through each segment (zone). For example, if they are successful in breaking through one of them, then the odds of them going further in are greatly lowered, given all of the layers of MFA that they have to go through. A simple example of the ZTF is in Figure 6.1.

As one can see from the above diagram, each segment (zone) has its own layer of MFA. But, also, it should be noted that there is no interlinkages that exist between them. The basic motto of ZTF is as follows:

> Never Trust, Always Verify

In this regard, it is expected that biometrics will play a much heavier role in terms of the MFA authentication mechanisms. For example, it will be used as least as one authentication mechanism, along with others, as reviewed so far in this book. It is also quite conceivable that it could be used for all of them. This is illustrated in the diagram in Figure 6.2.

As can be seen in the illustration above, there are three differing Biometric Modalities being used, in order to confirm the identity of the end user that is trying to gain access to the resources in the Shared Server. It could also be the case that even a mixture of both physical and behavioral biometrics

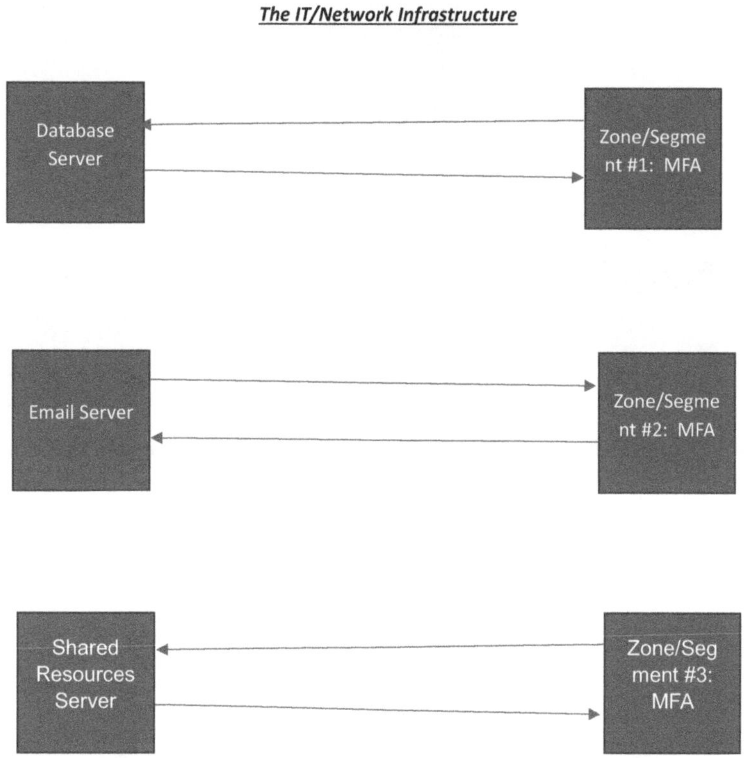


*Figure 6.1* This is an illustration of the zero trust framework.

can used as well, for the authentication mechanisms. This is also illustrated in Figure 6.3.

In this particular scenario, it is Signature Recognition and Keystroke Recognition that are the two behavioral biometrics that are being used. Finally, it should be noted that these kinds of methodologies just illustrated can be deployed for either an On Premises or Virtual ZTF.

## Edge computing

Another new trend that is starting to take shape is known what as "Edge Computing". The technical definition for it is as follows:

> Edge computing is an emerging computing paradigm which refers to a range of networks and devices at or near the user. Edge is about processing data closer to where it's being generated, enabling processing at

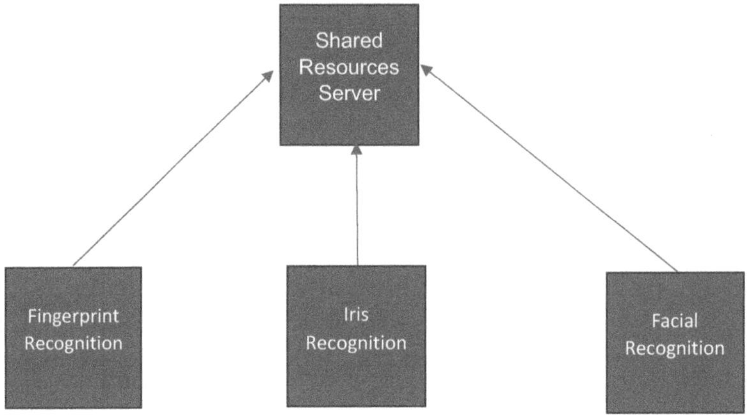


*Figure 6.2* This is an illustration of the zero trust framework, using biometrics for the MFA.

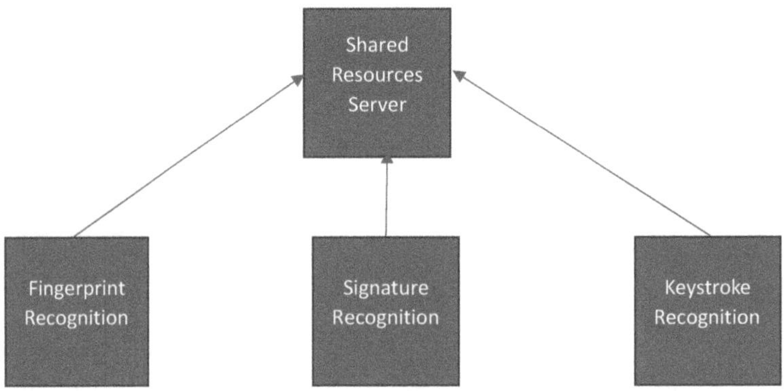


*Figure 6.3* This is an illustration of the zero trust framework, using differing Biometrics for the MFA.

> greater speeds and volumes, leading to greater action-led results in real time.
>
> *Source: Edge Computing | Accenture*

In other words, rather than having all of the data being stored and processed at a Central Server location, it is actually being done closer to the device of the end user. These two scenarios are illustrated in Figures 6.4 and 6.5.

In the first illustration, this depicts how the data is stored and accessed directly via the Shared Resources Server. The second illustration now demonstrates that while the data is still stored at the Shared Resources Server, it is actually being processed by a computer that is located within

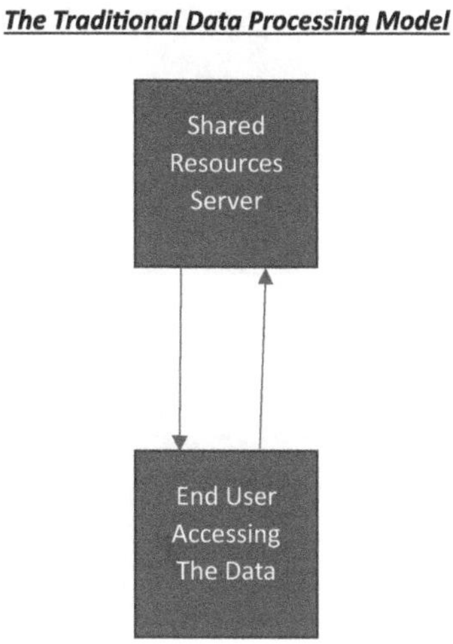


*Figure 6.4* This is an illustration of edge computing.

very close proximity to the device of the end user. This second scenario beings a number of key advantages which are as follows:

1. *Costs*:
   With edge computing, since the processing of the data and transmission of the data is done closer to the device of the end user, the costs of sending over large amounts of data and processing are much lower, as opposed to having them being done in a further away location, such as the Shared Resources Server.
2. *Latency*:
   This term can be technically defined as follows:

   Data latency refers to the delay or lag between the time when data is generated or sent and the time when it's received or processed. It can be likened to the time it takes for a message to travel from one place to another.

   *Source: What is Data Latency | GigaSpaces*

   This is illustrated in the examples in Figures 6.6 and 6.7.
   The first illustration shows the longer time lag that is experienced when the all of the data storage processing takes place at the location of the Shared Resources Server, whereas the second illustration shows the shorter time lag that is experienced when the processing is done closer to the proximity of the device of the end user.

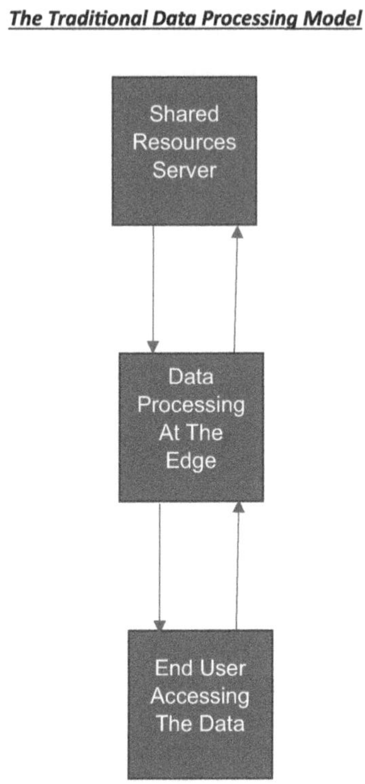


*Figure 6.5* This is an illustration of edge computing, where all of the data processing occurs closer to the device of the end user.

3. *Scalability*:
   By having the data processing done closer to the device of the end user, scalability can be achieved. This simply means that the overall data processing topologies can be quickly modified to meet the needs of the business requirements. This simply involves adding or subtracting the Edge Devices as needed.
4. *Redundancy*:
   Under the traditional methodology, where all of the data storage and processing occurs at the location of the Shared Resources Server, if there are any hiccups that occur within the network lines of communications from that to the device of the end user, or if the Shared Resources Server fails for some reason, it can take a longer time for the processes to resume to a sense of normalcy. But with edge computing, if the data processing occurs closer to the device of the end user, this gives a layer of redundancy, as the failure (if there are any), can be rolled over to yet another edge computing device,

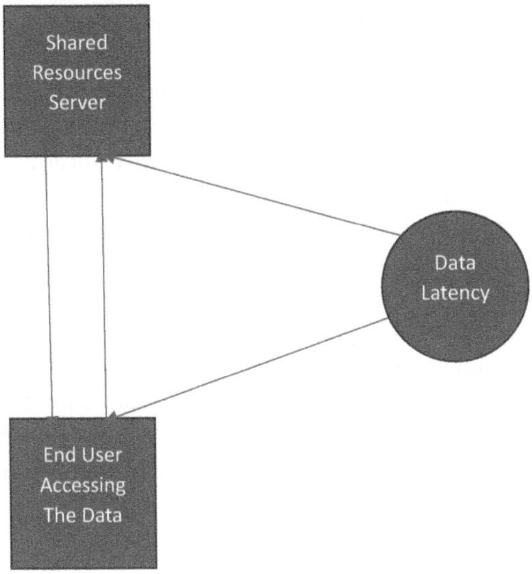


*Figure 6.6* This is an illustration of a shirt lag time, or shorter Data Latency.

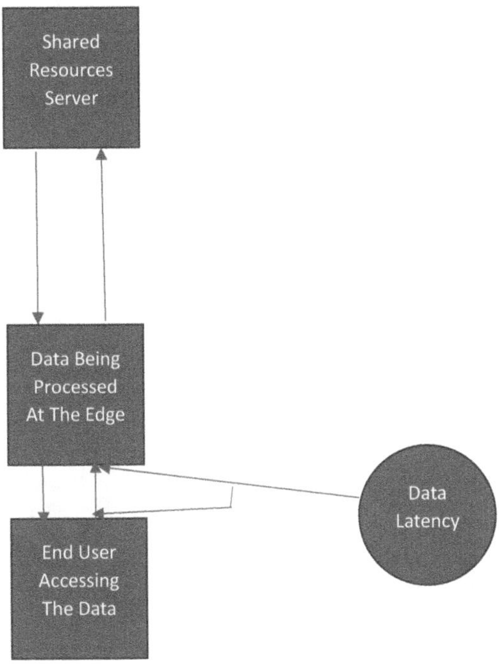


*Figure 6.7* This is an illustration of a short lag time, or shorter Data Latency.

assuming that there is another lines of network communications to the same Shared Resources Server where the data is being stored at.

5. *Real Time*:
   With the data being processed closer to the device of the end user, the data can be accessed on a much more real time basis, due to the fact that a much shorter lag time is being experienced.
6. *Attack Surface*:
   Since the processing of the data is occurring closer to end device of the user, the attack surface is actually greatly reduced. This simply means that the cyberattacker has a much narrower window in which to try to intercept the data that is being transmitted from the edge computing device to that of the end user. But, it is also important to keep in mind that data that is stored in the Shared Resources Server must also have the appropriate controls deployed onto it in order to mitigate the risks of any incidents data leakages occurring to it.

In the case of edge computing, Biometric Technology would be used to provide a means of MFA. This would occur at the point between the edge computing device and that of the end user. This is illustrated in the diagram in Figure 6.8.

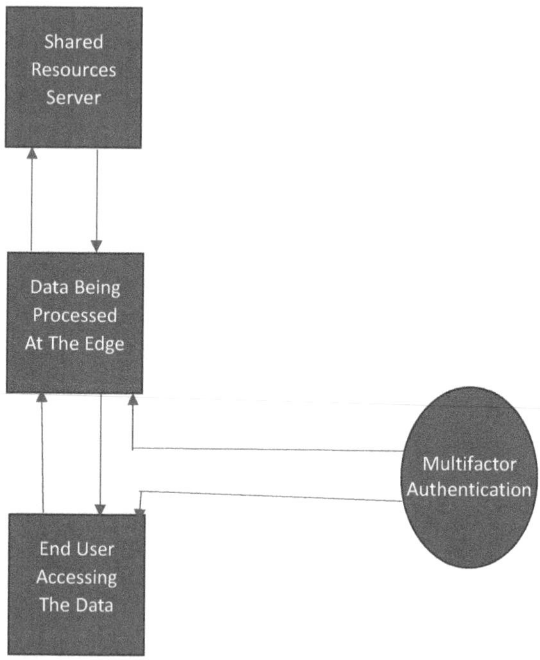


*Figure 6.8* This is an illustration of MFA being used in being used in edge computing.

*Figure 6.9* This is an illustration of an edge computing device.

In this scenario, the end user would be authenticated to the edge computing device by at least three or more differing mechanisms, one (or even all) which include a Biometric Modality. Finally, an illustration of an edge computing device being connected to the devices of end users (and vice versa) is illustrated in Figure 6.9.

## The Internet of Things

There are a few important terms that need to be defined, and they are as follows. The Internet of Things, also known as the "IoT", can be technically defined as follows:

> The Internet of Things (IoT) refers to a network of physical devices, vehicles, appliances, and other physical objects that are embedded with sensors, software, and network connectivity, allowing them to collect and share data.
>
> *Source: What is the Internet of Things (IoT)? | IBM*

As it can be inferred from the above definition, the IoT are all of those physical and even virtually based objects that are all interlinked together, through various lines of network communications. When one thinks of the IoT, very often the image of devices that can be found at an electronic store very often comes to mind. While this is true to a large extent, IoT devices can be also found in quite heavy usage in both the commercial and industrial sectors. This term is referred to as the Industrial Internet of Things, also known as the "IIoT" for short. This can be technically defined as follows:

> Industrial IoT is an ecosystem of devices, sensors, applications, and associated networking equipment that work together to collect, monitor, and analyze data from industrial operations.
>
> *Source: What Is Industrial IoT (IIoT)? – Cisco*

In simpler terms, this is where all of the devices in both the physical and virtual worlds are all interlinked among one another, via many network lines of communications. But instead of the applications being served for private and consumer uses, it is used by such industries as the Critical Infrastructure, Supply and Logistics Chains, Manufacturing, etc. But in terms of consumer usage, probably the largest known application of it is what is known as the "Smart Home". This too can be technically defined as follows:

> A smart home is a living space equipped with advanced technology that enables automation, remote control and real-time monitoring of various home systems and appliances. A smart home could, for example, allow you to control your thermostat, lights and security system with just a few taps on your smartphone – whether you're on the couch or on your way to work.
>
> *Source: Smart Home: What It Is, How It Works, Pros & Cons | Chase*

Put in another way, the Smart Home can be considered as a "watered down" version of both the IoT and the IIoT, because the network lines of communications that are interlinking all of the devices together are much simpler. For example, if you want to start a kitchen appliance, all one has to do is merely as the Virtual Personal Assistant (also known as the "VPA"), to do the needed task, such starting the brewing process for a cup of coffee.

Just like it has been reviewed in the last two subsections, it is highly expected that Biometric Technology will play a very large role all of these aspects of the IoT, especially when it comes to the standpoint of confirming the identity of the end user and authorizing them access to any device that is IoT or IIoT related. Below are some applications in which the role of Biometric Technology will play in this regard:

1. *Mobile Security*:
   Biometrics is already making a splash when it comes to Mobile Security. For example, for those end user who own an iOS device, the use of Facial Recognition (known technically as the "FaceID")

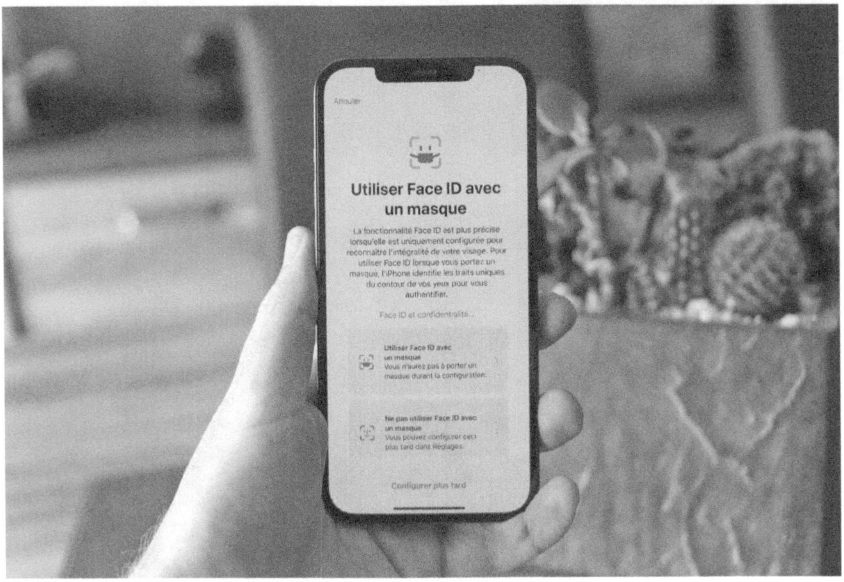


*Figure 6.10* This is an illustration of FaceID. (A person holding a cell phone in their hand photo – Free Iphone Image on Unsplash.)

and Fingerprint Recognition (known technically as "TouchID") are used as either as a Two Factor Authentication (2FA) or multifactor authentication (MFA) mechanisms. An illustration of the FaceID can be seen in Figure 6.10.

An example of the TouchID is in Figure 6.11.

In this regard, the use of a Mobile Wallet is an also considered to be an IoT component as well, as it interfaces with other payment systems, whether online or physical (such as a point-of-sale terminal). A Mobile Wallet can be technically defined as follows:

> A mobile wallet is a type of virtual wallet that stores credit card numbers, debit card numbers, and loyalty card numbers. It is accessible through an app installed on a mobile device, such as a smartphone or tablet.
>
> *Source: Mobile Wallet – Definition, Types, How To Use, Pros, Cons*

An example of a Mobile Wallet is illustrated in Figure 6.12.

As one can see from the illustration, all an end user has to do is merely tap their smartphone onto a point-of-sale terminal, and from there, the payment is instantly made. Even here, biometrics is heavily used as either a 2FA or MFA authentication mechanism.

*Figure 6.11* This is an illustration of TouchID. (Person holding black android smartphone photo – Free Apple Image on Unsplash.)

2. *The Smart Car*:
   Gone are the days of analog cars, where primarily the instrument panel was all just dials and needles. Today, the Smart Car has evolved, and while this can be considered a larger version of an IoT device, from within this, there are even other smaller IoT devices that exist as well. A technical definition of a Smart Car is as follows:

At the heart of the smart vehicle revolution is the integration of sophisticated electronics, sensors, and software.

*Source: What is a smart vehicle? – IOT Insider*

*Figure 6.12* This is an illustration of a Mobile Wallet. (Person holding black android smartphone photo – Free Payment Image on Unsplash.)

The IoT components that reside from within the Smart Car include the following:

- The collaboration to collect data.
- Autonomously adjust the vehicle's operations.
- Automatically adjust the maintenance, and comfort settings, thus greatly reducing the need for human intervention.

Some other, more robust features of the Smart Car include the following:

- *Autonomous Driving*:
  At the crux of Smart Vehicle innovation is what is known as "Autonomous Driving". This combines Artificial Intelligence (AI), machine learning, and other types and kinds of sensors which include the LiDAR, radar, and kinds of CCTV cameras. The end result is that these various IoT technologies empower smart vehicles to "understand" their surroundings and navigate to the most optimal route with little to no human oversight.
- *Connectivity*:
  Smart Cars possess state-of-the-art connectivity that integrates effortlessly with iOS and Android devices, wearable technology, and other

*Figure 6.13* This is an illustration of biometrics in a Smart Car. (Driver Warm Black Jacket Uses Keyless Stock Photo 2084886361 | Shutterstock.)

Smart Home systems. This allows for remote vehicular control, real traffic updates, and warnings about maintenance notifications.

Biometric Technology is not only used in Smart Cars to fully confirm the identity of the driver, but it is also even being used as a source for the ignition system of the Smart Car. For example, rather than using the traditional method of inserting the key into the ignition, a Fingerprint Recognition Sensor is now used. So with the quick swipe of the finger, the Smart Car will automatically start for the driver. This is actually a good security protocol, as only the owner of the Smart Car will be able to gain full access to it. An illustration of this can be seen in Figure 6.13.

But despite all of these key advantages that Biometric Technology brings to both IoT and IIoT devices, there is one grave cybersecurity risk: that is, the more interconnectivity that exists, the greater the attack surface expands for the cyberattacker to penetrate into. For example, through just one small backdoor in an electronic sensor, the cyberattacker can take control very easily of the entire Smart Car.

## Liveness detection

As it was reviewed in close detail throughout this book, whenever an end user submits a sample of either their physiological and/or behavioral traits, these typically get converted into a mathematical or statistical file format. This is then used to create the Enrollment Template (this is what gets stored

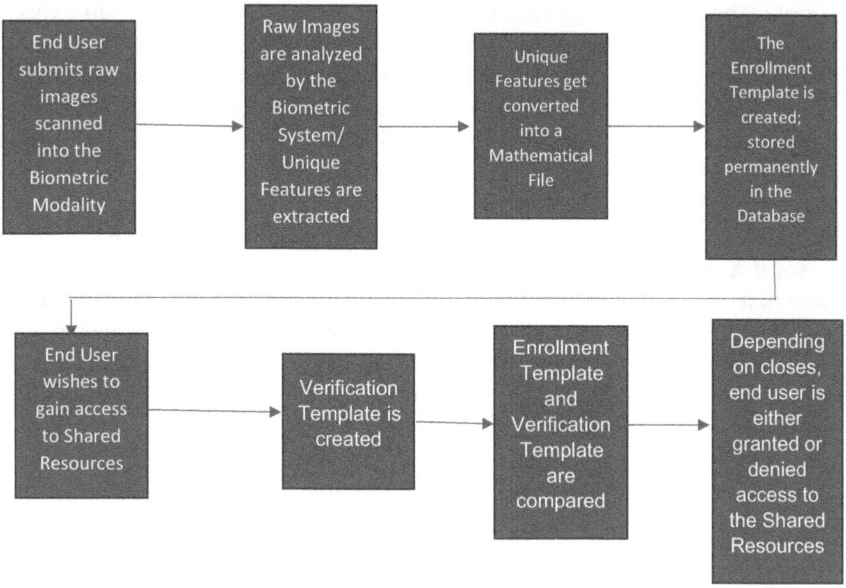


*Figure 6.14* This is an illustration of the Biometric Process.

permanently into the database of the Biometric System), and the subsequent Verification Template (which is not stored in the database of the Biometric System).

These two templates are then compared together, and depending upon the level of the closeness of the them, the end user is then 100% identified, and then allowed to gain access to the shared resources that their seeking to use. This process can be seen in the illustration in Figure 6.14.

To further illustrate this, take the example of Fingerprint Recognition. The end user submits their finger onto the sensor of the Biometric System, raw images are then captured of it, these are then compiled into one master raw image. From here, the unique features are then extracted, and they then get converted into a binary mathematical file format, which is nothing but a series of zero's and one's (11000011110001111). This becomes the Enrollment Template, which is stored permanently into the database of the Biometric System.

Once the end user wishes to gain access to the Shared Resources, they then submit their fingerprint onto the sensor of the Biometric System. This also becomes a binary mathematical file, just like in the Enrollment Template. The two are then compared, and depending on the degree of closeness of the two the end user is then either granted or denied access to the Shared Resources that they are seeking to use.

The question now often gets asked is this: "Can the Biometric Template be spoofed, or reverse engineered back into the actual, raw image?" On a theoretical sense, the answer to this is no. The Biometric Templates are once again, just mathematical or statistical representations of those collected raw images. A good analogy to this is that of Credit Card Theft. If the cyberattacker gains access to this, sure, they make all kinds of fraudulent purchases. But, if the cyberattacker were to hijack a Biometric Template, there is very little that they can do with it. The primary reason for this is that each Biometric Template that is created is unique to that algorithm that created it.

So, if a Biometric Template were to be stolen, the cyberattacker cannot take that an insert into a different Biometric System, because the algorithms will be different. But now, the caveat is that with the birth of Generative AI, this could potentially change the entire landscape on this. This has given to a huge uptick into what is known as the "Biometric Spoofing". A technical definition of it is as follows:

> Biometric spoofing is the deliberate manipulation or imitation of biometric traits to deceive systems reliant on biometric authentication for security. By using fake or altered biometric samples, attackers can pass off as someone else and trick authentication measures into giving them access to protected systems.
>
> *Source: Biometric spoofing definition – Glossary | NordVPN*

Now given the powerful Neural Network and machine learning Algorithms that partially drive Generative AI, it is now more likely than ever before that the mathematical files that are associated with both the Enrollment and Verification Templates could so how be reversed, but this is still an unlikely scenario. The more likely one is that the use of what are known as "Deepfakes" will now be one of the primary threat vectors in the cybersecurity Threat Landscape of both today and tomorrow.
Deepfakes can be technically defined as follows:

> A deepfake is an artificial image or video (a series of images) generated by a special kind of machine learning called "deep" learning (hence the name).
>
> *Source: What the heck is a deepfake? | UVA Information Security*

A sub field of machine learning is called "Deep Learning". Also, a technical definition of it is as follows:

> Deep learning is a subset of machine learning that uses multilayered neural networks, called deep neural networks, to simulate the complex decision-making power of the human brain.
>
> *Source: What Is Deep Learning? | IBM*

While is out of the scope of this book to review all of the mechanics of what Deep Learning is all about, it is still important to know that this is the primary vehicle that is used to create Deepfakes. A prime example of a Deepfake is during the Presidential Election Season here in the United States. For instance, when it comes closer to the time of the actual voting process, a Deepfake video can be created of one of the two major candidates. From here, the video can then be posted on all of the Social Media Platforms, primarily that of YouTube. In the Deepfake video, the fake representation of the candidate could be asking for money to donate to their campaign.

Once the victim is lured into this particular Phishing tactic, the donated money will not go to the actual campaign, but instead to a phony, offshore account, most likely a nation state threat actor, like North Korea, China, or Russia. An example of a Deepfake can be seen in the illustration below.

Although it can be very difficult to discern between the differences of a real image and a Deepfake one, there are some very subtle differences upon closer inspection. For example, in the illustration above, there are slight differences in the lighting, and both the eye and lip structures. So in actuality, this is now going to be a great boon for Biometrics Technology and, thus, is expected to be one of the biggest trends in the coming years.

For example, Facial Recognition can be used to with great accuracy to confirm the differences between a Deepfake video or image and that of the real one. The technical name for this new trend is called "Liveness Detection", and it too can be defined as follows:

> Liveness detection, or a liveness check, is a security method that ensures an entity presenting a biometric, like a face or fingerprint, is a live human being and not a fake image or recording.
>
> *Source: What Is Liveness Detection? How ItHelps Fraud Prevention*

There are three distinct ways in which liveness detection can work, and they are as follows:

1. *Motion Analysis*:
   The movements of the image or video in question are being examined for natural and successive motions and behavior patterns that a live human would produce.
2. *Three Dimensional Sensing*:
   This is also technically known as "3D Depth Sensing". It makes heavy usage of laser-scanners-related technologies to map out a person's facial features into various three-dimension models.
3. *Texture Analysis*:
   This technique is used to take a closer look at the biometric features, particularly those in the facial area. This is typically done to detect

signs of life that are typically not noticed accurately in a photo or video. Examples of this include the following:

- Fine lines wrinkles
- The shape of the pores
- The texture of the skin
- Other imperfections or "rough areas"

In conclusion, this concluding chapter has examined what some of the top trends will be for Biometric Technology in 2025 and beyond.

Overall, this book so far has covered the following:

1. Chapter 1:
    - The various issues surrounding Biometric Technology, which include the following:
    - The societal impacts of biometrics between both the developed and developing nations.
    - Some of the major market applications for biometrics.
    - A review of how security documents are protected by biometrics.
    - How developing nations favor the use of biometrics much more so than the developed nations.
    - The issue of privacy rights as it relates to biometrics.
    - The fear of identity theft.
    - The ergonomic factor in the biometric modalities.
    - The phenomenon of function creep.
    - The legal considerations of Biometric Technology and its respective uses.
    - The legal status of a Biometric Template.
    - The storage of the Biometric Templates.
    - The usage of smart card technology.
    - The legal admissability of Biometric Templates in a court proceeding/trial.
    - How biometrics are used in e-voting.
    - A technical review of the e-passport.
2. Chapter 2:
    - The Biometric Technologies of today – physical and behavioral.
    - The differences between physical biometrics and behavioral biometrics.
    - Which one to use? Physical biometric or behavioral biometric?
    - A review of fingerprint recognition.
    - A review of hand geometry recognition.
    - A review of vein pattern recognition.
    - A review of palm print recognition.
    - A review of facial recognition.
    - A review of iris and retinal recognition.

- A review of voice recognition.
- A review of signature recognition.
- A review of keystroke recognition.
- A review of DNA recognition.
- A review of gait recognition.
- A review of earlobe recognition.

3. Chapter 3:
   - Biometric technology system architecture.
   - Sensing and data acquisition.
   - Multimodal biometric systems.
   - Single sign on solutions.
   - Implementing a multimodal biometric system.
   - Challenges with multimodal biometric systems.
   - Signal and image processing.
   - Preprocessing of the biometric raw image.
   - Quality control checks.
   - Feature extraction.
   - Data compression.
   - Data storage.
   - Search and retrieval techniques.
   - Database search algorithms.
   - Backup and recovery of the database.
   - Database configurations.
   - Template matching.
   - Threshold decision-making.
   - Establishment of the security threshold values.
   - Data transmission.
   - Biometrics project management.
   - System concepts and classification schemes
   - The biometric interface
   - Biometric networking topologies
   - Network protocols
4. Chapter 4:
   - Introduction to cryptography
   - Message scrambling and descrambling
   - Encryption and decryption
   - Ciphertexts
   - Symmetric key systems and asymmetric key systems
   - Polyalphabetic encryption
   - Block ciphers
   - Cipher block chaining
   - Disadvantages of symmetric key cryptography
   - The key distribution center
   - Mathematical algorithms with symmetric cryptography

- Asymmetric key cryptography
- Keys and public private keys
- The differences between asymmetric and symmetric crytography
- The disadvantages of asymmetric cryptography
- The mathematical algorithms of asymmetric cryptography
- The public key infrastructure
- The digital certificates
- How public key infrastructure works
- The public cryptography standards
- Parameters of public keys and private keys
- Securing the public keys and the private keys
- Virtual private networks
- Mobile virtual private networks
- Conducting a virtual private network cost benefit analysis
- Implementing a virtual private network
- Virtual private network testing
- Managing public and private key exchanges
- Four vulnerabilities of a biometrics virtual private network
- Biocryptography
- The cipher biometric template
- Biocryptography keys
- Biocryptography and virtual private networks
- IPSec

5. Chapter 5:
   - Introduction to cloud computing
   - The basic concepts and terminology surrounding cloud computing
   - The cloud
   - Scalability
   - Availability and reliability
   - The challenges and risks of cloud computing
   - Reduced operational governance
   - Compliance and legal issues
   - The functions and characteristics of cloud computing
   - On demand usage
   - Resource pooling
   - Elasticity
   - Measured usage
   - Resiliency
   - Cloud computing delivery models
   - PaaS
   - SaaS
   - Cloud computing deployment models
   - Public cloud

- Private cloud
- Hybrid cloud
- The security threats posed to cloud computing
- Important mechanisms of cloud computing
- Load balancers
- Cloud computing cost metrics and service quality mechanisms
- Service performance metrics
- An introduction to biometrics in the cloud
- A detailed example of how biometrics in the cloud would look like
- A review of the advantages and the disadvantages of biometrics in the cloud
- The AWS and Microsoft Azure, and how they fit into biometrics in the cloud.

# Index